THE
GREAT APES

THE GREAT APES

Our Face in Nature's Mirror

Text and photographs
Michael Leach

BLANDFORD

ACKNOWLEDGEMENTS

A long list of people have helped in the production of this book – too many to acknowledge in full. However, I would like to say a particular thank-you to Ashley Leiman of the Orangutan Foundation, Greg Cummings of the Dian Fossey Gorilla Fund (UK), and Nick Ellerton and Beryl Ramsay of Chester Zoo. Linda Offord did an excellent job in picking up last-minute mistakes in the text. Thanks are also due to the late Barklay Hastings for the use of his picture of Jozi on p. 149.

I would also like to add a special note of appreciation to Pentax UK Ltd and John Dickens for help with camera equipment. All of the photographs in the book were taken on a Pentax Z1. Conditions were frequently brutal, but camera and lenses performed unbelievably well, even after being beaten up by an orang-utan.

I would like to give heartfelt thanks to my wife, Judith, for reading through the text and correcting errors that still make me wince.

A BLANDFORD BOOK

First published in the UK by Blandford
A Cassell Imprint
Cassell Plc, Wellington House,
125 Strand, London WC2R OBB

Copyright © Michael Leach 1996

Distributed in the United States by Sterling Publishing Co., Inc.,
387 Park Avenue South, New York, NY 10016–8810

Distributed in Australia by Capricorn Link (Australia) Pty Ltd
2/13 Carrington Road, Castle Hill, NSW 2154

British Library Cataloguing-in-Publication Data
A catalogue entry for this title is available from the British Library

ISBN 0–7137–2488–9

Typeset by Keystroke, Jacaranda Lodge, Wolverhampton

Printed and bound in Hong Kong

CONTENTS

PREFACE

As you sit and read, the subjects of this book and our closest relatives are slowly and quietly sliding towards extinction. We are, through thoughtlessness and greed, destroying the habitat that is essential for their survival. Unless we act immediately, their complete disappearance is probably a matter of decades rather than centuries away.

The great apes are confined to the tropical areas of Africa and South-East Asia, regions where logging, mining and hunting threaten their very existence. At the close of the twentieth century, it is astonishing that we have still not yet learned to take responsibility for our actions. We can build super-intelligent computers and send probes to beyond the outer edges of the solar system, but in spite of all we know, humans continue to behave like bulls in a china shop. We inflict terrible damage on the earth and its residents, and we lack the power to rectify the long-term effects of our short-term profligacy.

In the wake of such carelessness comes the vacuum of extinction. Over thousands of years humans have notched up the destruction of countless animal and plant species which will never again be seen, all because we did not think and act to save them. We are no better today. The International Union for the Conservation of Nature and Natural Resources publishes the *Red Data Book*, which lists all of those species that are now in real danger of extinction. Not all are threatened by our actions, but the vast majority are in trouble because of us. Among them are the four great apes: gorillas, orang-utans, chimpanzees and bonobos.

Most people only ever have the chance to see apes in zoos. Here they are often regarded as clownish, subhuman creatures not quite worthy of the awe and respect reserved for a big cat or shark. But these institutionalized animals bear little resemblance to their wild counterparts. The traditional 'popular' view of apes was long ago proved to be false and misleading in almost every way, but deep-seated myths are difficult to sweep aside.

In their native forests, the great apes are supreme masters with few enemies capable of tackling them. The cuddly tea-drinking chimp is potentially one of the world's most dangerous animals. They hunt efficiently and lethally in packs, each ape with a specific task and position, in order to catch, kill and eat monkeys. The gorillas' reputation for blood-curdling ferocity is completely unrecognizable to those fortunate travellers who

The name 'common chimpanzee' is highly misleading, for these apes are becoming increasingly endangered in the wild.

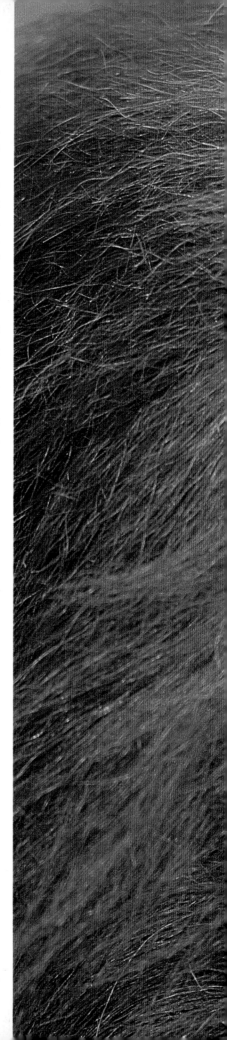

have had fully grown wild gorillas sitting on their knees. Until very recently orang-utans were thought to be the slothful, intellectual weak links of the family. Now we know that their brains, though different, are just as keen as those of the other apes.

To see a wild ape in its own environment is to look at ourselves as we might have been. High in the Virunga Mountains of Rwanda, I once sat in a rain forest surrounded by mountain gorillas. The huge silverback male stood aloof and alert about 5m (16ft) away, but the females and youngsters felt no such reserve. One two-year-old was exploring my camera bag, while another chewed a bamboo shoot less than an arm's reach away. An adult female pushed her way through a bank of giant nettles and sat directly in front of me. She stared; nothing else. Her hazel-brown eyes studied my face and clothes. The gorilla showed no signs of fear or aggression, just a lively interest in a visitor. It occurred to me that exactly the same scene would take place if I were introduced to a stranger at a party. She was weighing me up. After a minute or so she gently brushed my bare arm with the back of her hand, sniffed her fingers to investigate the scent and then wandered off to eat. She had used four of her senses to gather information about the newcomer and was now satisfied. I have watched wildlife all over the world but nothing has ever affected me as much as this brief meeting.

Animals generally see humans as an enemy to avoid. In order to study them, we often have to use secretive devices such as hides or binoculars, keeping us detached and remote from the animals. But this mountain gorilla chose to approach me; and not out of belligerence or hunger, the two stimuli which most often initiate close encounters between humans and animals. There was no ulterior motive to her actions, she was merely curious. Looking at the gorilla was like staring at a distorted fairground mirror. The reflection was not quite mine but it was so very close. The overall appearance and features were almost familiar. I had no doubt that this was the face of a long-lost relative.

The great apes are one of the small number of animals that often disturb humans on a very basic level. As a rule apes do not frighten people; they make us uncomfortable. We generally categorize animals into convenient, non-zoological groups so that we can relate to them. Some species, such as the great white shark and the tiger, are regarded with reverence and/or fear for their overwhelming speed and power. Garden birds are welcome visitors that we like to feed through the cold months of winter, for they allow modern man to keep in contact with nature without too much difficulty or trouble. Domestic animals are bred for our convenience in the form of company, transport or food. Other species invoke admiration for their beauty, agility and countless other

Orang-utans have the reputation of being the least intelligent of the apes, but those who work with them know better.

traits that we find interesting. But apes cannot be pigeon-holed so readily. For many people they are, in every way, too close for comfort.

In the time taken to write this book, I have bumped into many people who, having discovered the subject of the project, responded along the lines of, 'How can you bear to work with those animals? They are so spooky/sinister/weird/strange.' This has come to be known as the ape complex and is a manifestation of the unease we feel when looking at our closest relatives. An awful lot of people don't like being reminded of the fact that man and animals are not really that different. The eighteenth-century biologist Lorenz Oken, a man who even then should have had a more detached view, noted:

Apes resemble man only in their bad habits and behaviour. They are wicked and deceitful, spiteful and indecent, adept at learning tricks, but disobedient, often breaking up a game by foolery . . . like a clumsy clown. There is not one virtue which man can admit the apes to possess, still yet a use they could have for man . . . They are merely the bad side of mankind, from the physical as well as the moral viewpoint.

From a purely technical point of view, writing a book on any of the higher mammals is a tangled matter. With simple life forms such as the ant, behaviour is purely instinctive and rarely deviates from the straight and narrow. They simply do not have the mental capacity to experiment with other possibilities. But apes, like humans, show behavioural patterns that are a complex combination of pre-ordained genetic coding and personal temperament. Just as individual humans vary in intelligence, levels of curiosity and bravado, so do apes. The behaviour of each of us is uniquely moulded by our own personal experience combined with the natural skills we are born with. Apes are shaped by the same factors and this makes it difficult to lay down precise rules for their behaviour.

It is possible to work out a set of circumstances that might be met by an ant colony and accurately predict how the ants will react in advance, for their responses are rigidly set. The same cannot be said for humans – or apes. Both are highly intelligent groups with an astonishing potential for individual variation in behaviour and abilities. An inevitable result of high intellectual power is the capacity for learning and rapid adaptation. Humans have become so powerful because we are able to modify our actions so readily.

Although great apes do not show the same talent for such rapid change, it would be foolish to underestimate their learning skills. When it comes to predicting possible patterns, all we can do is state how other apes have been seen to act before under similar conditions. Every year there are reports of new behavioural traits that have never been previously recorded. Where populations are isolated from the rest of their species, the whole group may have adopted habits that differ slightly from those of

Popular mythology portrays these animals as bloodthirsty killers, but in reality gorillas are gentle and playful vegetarians.

their distant cousins. Bushmen in South Africa and insurance brokers in Manhattan are the same species, but their lifestyles and behaviour are worlds apart.

This book is a guide to the great apes, not an unbreakable template for their behaviour. It would be possible to write a massive volume on a dozen of the different subjects covered in each of the following chapters. And even that would still leave out most of our accumulated knowledge. Apes are a fascinating group *because* they are so complex. It is hard enough for us to understand the workings of another human's mind. How much more difficult is it to analyse the actions and thoughts taking place in the brain of a totally different species – particularly when the animals involved are as individual and variable as the great apes.

I would find it difficult, if not impossible, to write a book on a subject that did not interest me. As a full-time naturalist, I obviously have some instinctive feeling for the natural world, but even for a hard-bitten professional like me, apes are special. My first contact with a live specimen came when I was working as a temporary zoo keeper before starting college. One morning we were informed that the zoo was taking charge of a young chimpanzee who had been kept as a pet but had grown out of control. We were warned that this six-year-old female was a real hand-ful. With great care the strong wooden box was unloaded and opened. At the back, on a thick bed of wood shavings, sat a badly frightened ape. With wide grins and held-out hands, she signalled her fear. Tempted by several bananas, the chimp was calmed and eventually sat on my lap. So, I thought, this is the tyrant that terrorized a family home. I looked into her bright, intelligent eyes and the chimp, now relaxed and inquisitive, put an arm around my neck, leant forward and bit me through the ear. She then did her best to wreck the ape-house kitchen. It was a lesson I was never to forget: apes are powerful and formidable animals. Today I have grave doubts about the ethics of keeping these animals in zoos. Some, such as Jersey, San Diego and Plankendael in Belgium, do an excellent job for both conservation and education, but they have exceptional facilities with large enclosures and expert staff. Unfortunately, other zoos are less well equipped and the apes inevitably suffer to some degree.

Taking a long-term overview of the earth and its resources, the great apes are no more – or less – important than any other group of animals. But maybe they do deserve a little extra attention occasionally. For a variety of reasons all are threatened to some degree, and some have their backs to the wall in a last-ditch stand. It will take only a gentle nudge to push the mountain gorilla over the edge and into the oblivion of extinc-tion. Humans simply do not have the right to commit such an appalling crime against the future of the planet.

On a more personal level, apes are our closest living relatives. How can we sit by and watch them die out before we even really start to understand

The few travellers that are fortunate enough to see gorillas in the wild find animals that are shy and gentle. They bear little resemblance to the violent gorillas of popular myth.

It is the face of an ape that both attracts and disturbs many people. After humans, chimpanzees have the most expressive faces of any living animal.

them? The apes' predicament is entirely of our making and only we have the power to rectify the situation. I have worked with dramatic and exciting animals all over the world, but the mountain gorillas made more of an impact on me than any other species. My children and grandchildren may never have the chance to sit in a forest surrounded by gorillas, but I would like them to know that they live in a world where these animals still exist. I cannot add anything to the millions of words already written about the obscene tragedy of extinction, I can only add another voice. Let's help now, before there are no apes left to help.

Michael Leach, Shropshire

1
A SHARED
ANCESTRY

THERE IS AN OFTEN repeated story about the wife of a nineteenth-
century bishop of Worcester who, on hearing of Charles Darwin's
new-fangled theory of man's evolution, is reported to have
commented: 'My dear, descended from the apes! Let us hope it is not true,
but if it is, let us pray that it will not become generally known.' This is
a perfect illustration of the misunderstandings that have followed on from
Darwin's revolutionary thinking over the past 140 years. For at no stage
did he suggest that man was descended from the apes; he merely put forward the idea that perhaps we originally came from the same stock and,
in the dim and distant past, may have shared a common ancestor.

Darwin is credited as the instigator of modern evolutionary science. He
was certainly the first to put the idea into lucid prose and give it a good
scientific grounding, but it was not a completely new concept. Lucretius,
the first-century BC Roman poet, suggested that man's ancestors were
wild animal-like creatures who hunted with their bare hands and teeth
before slowly moving on to use pointed sticks. Galen, a Greek physician
living in the second century AD remarked that a monkey was 'most
similar to man in viscera, muscles, arteries, veins, nerves and in the form
of bones'. Leonardo da Vinci, while experimenting in the illegal dissection
of human corpses around the end of the fifteenth century, noticed undeniable similarities between the human and primate anatomy. He wrote:
'. . . describe man, including such species as are of almost the same
species, as Apes, Monkeys and the like'. Later he concluded: 'Man in fact
differs from animals only in his specific attributes.'

Each culture had its own explanation that covered the creation of the
earth and its residents; and all were inventions to cover a total void in
their knowledge of how life really started. For centuries the accepted
wisdom in the Western world was simply to take the Bible at face value,
including the story of Creation. It was a matter of faith that everything
was created by God exactly as laid down in Genesis. In 1650 James
Ussher, Archbishop of Armagh, announced that the world was created in

*Modern monkeys such as this Celebese macaque are closely related to both man and
the apes.*

4004 BC. He reached this date by carefully going back through the Old Testament and adding together all time references. The date was later confirmed by Dr John Lightfoot of St Catherine's College, Cambridge, who pinpointed the exact moment of creation as 9.00 a.m. on 23 October. Few people questioned this; after all, it was as reasonable an explanation as any other and it had as proof a detailed chronology in the Bible.

But irritating pieces of evidence kept cropping up that cast doubt on the biblical account. Huge bones were excavated that bore no resemblance to any living creature; there was no mention of such giants anywhere in the Bible and surely nothing that massive could possibly have been over-looked. A rational explanation soon come forward which unravelled the mysterious existence of these extraordinary fossils. Several theologians, championed by the French geologist Baron Georges Cuvier, argued that the animals living on earth, as we know them, now were not the first attempt at creation. There had been other earlier and less perfect worlds containing strange and misshapen creatures whose sinful ways had displeased God, making them undeserving of life. These species had been destroyed by global disasters on a scale similar to the great flood and it was their unfamiliar mortal remains that lay hidden in the earth. A team from the Académie des Sciences calculated there had been exactly twenty-seven successive creations; shortly afterwards William Smith, an Englishman, put the number at thirty-two. The familiar and accepted Genesis account told only of the last and most perfect creation, that which contained man. This idea was no more implausible than the others and it made sense given the few scraps of evidence available.

Fossil bones were viewed and dealt with in extraordinary ways. There was a fleeting school of thought asserting that man's early ancestors were a race of giants. In 1577 a collection of mammoth bones, as we now know them to be, were seriously going to be given a full burial service and inter-ment in a graveyard until it was decided that, as the giant 'man' came from a pre-Christian era, the ceremony would have been unsuitable. A century later, a Frenchman by the name of Henrion, basing his calculations on fossil evidence, announced that Adam was 38m (123ft 9in) tall while Eve was 36.5m (118ft 9in).

In 1726 a strange skeleton was unearthed on the shore of Lake Constance and was examined by several worthy authorities of the day. One such notable, Johann Scheuchzer, a Zurich doctor and canon, declared the find to be a preserved example of pre-Flood man: '. . . the bony skeleton of one of those infamous men whose sins brought upon the world the dire misfortune of the Deluge'. The good doctor then went on to say that this unfortunate fellow died in 2306 BC, the supposed year of the biblical Flood, and his strange appearance was the result of living and dying in appalling sin (this particular fossil has since been identified as a giant salamander from the Tertiary period). Long lines of new theories were fashionable for a while, before completely disappearing. For a time in the late eighteenth century, the orang-utan was seen as the mythical missing link that bridged the divide between man and animals. When the species was recognized as simply an Asian member of the ape family, the link idea was abandoned.

The 'ultimate creation' theory conveniently explained away monster fossils from beneath the ground, but in the mid-nineteenth century there was a growing number of scientists and philosophers who refused to accept its doubtful logic. Then in 1871 a book was published, written by an Englishman equipped with a degree in theology and the classics, which clearly and scientifically argued the case for true evolution. *The Descent of Man* had the audacity to dispute the biblical account and suggest that humans were actually descended from far more mundane roots. Charles Darwin's elegant and far-seeing theory opened up a whole new world of possibilities. Lacking the data we have today, Darwin's ideas were surprisingly sophisticated and accurate. He had no way of knowing that scattered around the world were fossilized bones, still buried, that would one day prove his theory by showing a series of subtle anatomical changes that connected long-extinct species via a chain of evolutionary stepping stones to animals that we all know today. His theory took both the religious and scientific communities by storm, and attracted world-wide interest from countless people who were not intrinsically involved in either camp. Darwin was simultaneously lionized and scorned, scorching attacks were published and the young scientist became the butt of much humour, both satirical and downright cruel. The man himself had doubts about challenging accepted wisdom and deeply held beliefs; he knew that these ideas would shake the establishment to its core. He wrote: 'At last gleams of light have come, and I am almost convinced . . . that species (it is like confessing to a murder) are not immutable.' Fortunately, from the beginning there were others who could recognize the inarguable sense behind his theory, and a small but significant group of supporters gathered behind Darwin. The debate and resulting furore captured the imagination of the press and public. Could modern humans really look into the past and see long-dead ancestors that ran on all fours and were covered with hair? With limited information and an open mind, Darwin dragged biological science into the modern era, but, as we see now, his theories barely scratched the surface of the mutual story lying behind the history of ape and human evolution. Over a century later, we are still trying to fit together the fragmentary pieces of a giant jigsaw. And there are huge chunks of the overall picture that have to be filled by guesswork.

The starting point of any evolutionary tree is always elusive, for if we go back far enough it is possible to show that every animal is related to every other animal. But the relationship is strained, to say the very least. Looking at primates, we need go back no further than the point immediately before present groups began to appear and go off in their own directions. Most modern zoologists classify apes into three distinct families: the Hylobatidae (gibbons or lesser apes), the Pongidae (gorillas, orang-utans and chimpanzees) and Hominidae (humans). Within these groupings there is a good deal of argument about the exact relationship between the species.

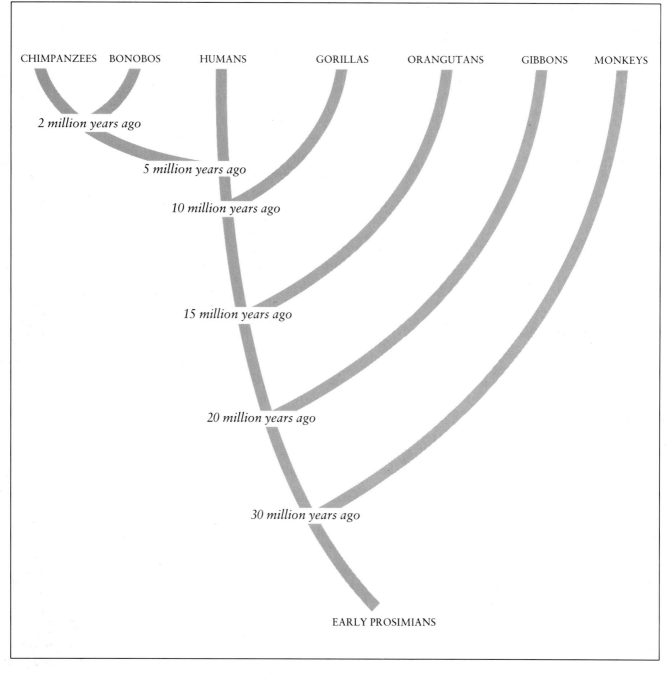

Fig 1
It is possible to draw many different family trees for primates. This is just one suggestion for the route taken by the ancestors of the modern species.

Ape-like creatures did not appear in any recognizable form until the Miocene period (22 to 14 million years ago). Around 50 million years ago there were huge numbers of species that are collectively known as the prosimians or pre-monkeys. These are well represented in the modern world by the tarsiers and lorises of Asia, the lemurs of Madagascar and the bushbabies of Africa. The prosimian group had a far greater distribution 50 million years ago than they have today, but as more efficient species appeared they were driven out and pushed into peripheral areas. Modern lemurs survive on Madagascar because the island was cut off from the rest of Africa by a channel of the Indian Ocean around 30 million years ago, isolating the lemurs against the irresistible tide of the more advanced primates that were appearing elsewhere. The lemurs had a further advantage in that Madagascar has no large resident carnivores to whittle their numbers down. The bushbabies and lorisies survived either because they became nocturnal to avoid competition or because they had already abandoned a diurnal lifestyle and the newer species, being creatures of the daylight, posed no direct threat.

Early monkeys evolved in the dense, warm forests that carpeted much of the tropics. Agile, climbing animals that fed in the canopy were the most suitable life form to exploit this environment and even in today's primary rain forests monkeys are the most diverse and numerous of all mammals. From an arboreal beginning, species later evolved that spent more time on the ground and eventually left the woodland in favour of bush or even open grassland. Familiar modern species of prosimians are probably quite similar to the common ancestors of apes and monkeys. They would have been small and agile with skilled, nimble fingers and brains that were very advanced for their time.

Prosimians, monkeys and apes together make up the order of primates. This is formed of twelve families, including man, covering a total of 182 known species. Prosimians differ from true monkeys by having a far better sense of smell and a smaller brain. They also tend to have longer, dog-like snouts while monkeys have more rounded heads, with the exception of baboons. Monkeys rely more on sight than on smell and their brains account for a greater percentage of their overall bodyweight.

Lemurs are the outstanding example of prosimians and their domination of Madagascar is an indication of just how successful our early ancestors must have been. There are ten recognized species, with many more known to have become extinct. Modern lemurs should not be seen as carbon copies of early prosimians, for they have evolved just as other animals have. However, lemurs do bear a strong anatomical resemblance to early primates and this suggests that their lifestyle was probably quite similar. The name 'lemur' means ghost and refers to their mysterious and once unknown life in the trees. Extinct lemurs were even more varied in size and appearance than they are now; one species was the size of an adult chimpanzee. An entire subfamily, known as the Megaladapinae, contained

All primates appear to have evolved from animals that resembled this slow loris from Asia.

animals the size of orang-utans which have entirely disappeared. Lemurs of today, however, are not just relics of the past waiting to vanish; they have evolved complex specializations that are comparable to those of monkeys. There are nocturnal and diurnal species; some feed on the ground and others forage in trees; some hunt insects and others eat leaves. This is a diverse and locally successful family of animals. Unfortunately, these cat-sized creatures are confined to just one small part of the world and, due to the inevitable interference of man, some species are under extreme threat. Humans arrived on Madagascar only 2,000 years ago, nearly 30 million years later than the first lemur-ancestors were isolated on the island. Hunting and habitat destruction have resulted in a terrible decimation of all known species. Lemurs are animals to be fully protected and studied, for it was creatures such as these that gave birth to the true primates.

Modern primates range in size from the tiny dwarf bushbaby, which weighs 100g (4oz), to the mountain gorilla, which is more than 1,200 times heavier. All species have several common factors that help contribute to their success. Primitive prosimians would have been insectivorous and equipped with claws for digging out grubs. Evolution

Ringed tails and other lemurs of Madagascar are a remnant population of the prosimians that eventually produced monkeys and apes.

dispensed with these in favour of flat nails on the end of each finger and toe. Nails combined with soft, sensitive fingertips provided nimble tools that could still be used to scratch and dig, but also possessed a fine tactile sense and manipulative skills that outshone those of any other animal. At some stage there was a gradual sideways shift of the inside digit. This finally moved 90 degrees away from the index finger and eventually resulted in the distinctive primate thumb (and the less useful human big toe). This seemingly trivial event was to be one of the turning points in our collective history, for the opposable digit turned a hand or foot from a flat, spade-like surface into a sophisticated device capable of gripping branches, peeling fruit, using tools and manipulating tiny objects. The thumb provided both a power grip when pure strength was needed and a precision grip for more delicate control. The importance of an opposable thumb is impossible to evaluate, but it is worth remembering that even chimpanzees, the most dextrous of all apes, cannot match the fine control that humans have over their hands. Chimps can pick up small objects only by holding them between their thumb and the first joint of their index finger. This awkward grip gives them excellent manipulative skills, but in the same circumstances a human would use the tip of a thumb against the tip of the finger, resulting in far greater precision.

Life in a forest canopy would have inevitably produced some animals that learned to make full use of the environment by mastering climbing and jumping skills. This would have led to a honing of stereoscopic vision. The ability to judge distance accurately is vital to an animal that lives in a complex three-dimensional environment such as a tree. Primates have forward-facing eyes and the right sort of brain to effortlessly calculate distances by combining two images to produce a single three-dimensional picture. They also needed to develop acute colour vision to help them pinpoint fruit among the overwhelmingly green backdrop of leaves. The advantages brought about by an opposable digit, such as the new-found ability to swing through trees, possibly explain the increase in brain size. Rapid and efficient intelligence is required to judge distance, speed and bodyweight, put the factors together and evaluate the possibilities of leaping from one branch to another – particularly when the branches are dozens of metres above the forest floor and an error is likely to mean death. The power of the early prosimian brain certainly had to expand to make full use of the animals' physical possibilities.

Today the list of monkeys extends to around 130 living species and many more are now extinct. The first true monkeys appeared around 35 million years ago, during the Oligocene period, and they spread rapidly. Twentieth-century monkeys are split into two groups: the Old World species from Africa and Asia and the New World species from South America. The most obvious difference between the two is nose structure.

New World monkeys have wide, splayed nostrils that are set far apart, while Old World animals have slender, closely set nostrils that generally face down. Their dentition also differs. Old World species have thirty-two teeth, while New World animals have four extra premolars. Another important characteristic is the adoption of a prehensile tail. Many New World species have evolved a highly flexible and tactile tail that is used as a fifth limb when climbing; this is completely unknown in Old World species. A fully prehensile tail has no hair on the underside; instead it has fleshy ridges that provide added grip when the monkey climbs. This, plus a strong and complex group of muscles, gives the monkey an extremely useful extra limb. Despite the differences, New and Old World groups are closely related; the anatomical contrasts appeared only after their ancestors were separated geographically.

Monkeys are a highly successful group that have evolved in wonderfully inventive ways. The smallest is the pygmy marmoset from South America, which weighs just 170g (6oz) and the largest is the African male mandrill, which can weigh up to 50kg (110lb). Generally, monkeys are fruit and leaf eaters, but many supplement their diets with animal protein in the form of insects, eggs and even small creatures such as young birds and lizards when given the chance. Monkeys and humans share many features that are essential for survival. Forward-facing eyes that pick up fine detail, judge distance and perceive colour are one of the primates' most valuable assets. The prosimians' sense of smell has been superseded in advanced primates by acute eyesight. This may be one reason why many primates have hair-free faces, while the rest of their bodies are covered in a thick coat. Many mammals communicate through scent, but as the primates' sense of smell declined, they adopted facial expressions as a visual alternative and so faces became bald for clearer signals. The expressive use of the face has been refined and taken to its ultimate with humans.

Monkeys are creatures of the daylight and all species but one are diurnal. The douracouli, or night monkey, is one of the few real specialists in the family. It is interesting to note that the only nocturnal monkey lives in South America, a continent where there are no prosimians. The only diurnal prosimians exist in Madagascar, an island with no monkeys or apes. Where the two exist side by side, monkeys and apes are diurnal and prosimians are nocturnal, thus avoiding direct competition. Other monkey specialists include the tiny marmosets and tamarins that still retain claws on all but the first digit of each hand. These cat-like hooks are used as grappling irons for climbing the huge tropical trees in which the animals search busily for fruit. The remaining finger is equipped with a flat nail which is used in exactly the same way as by other primates. Zoologically, marmosets may be classified as monkeys, but scurrying and scampering around the branches, they seem more like rodents than anything else.

Tails come in all shapes and sizes in the monkey world. As well as being used as fifth limbs by New World species, they are essential aids for all as they balance in the trees. Like an acrobat carrying a long pole when walking a high wire, the tail of an arboreal monkey is used as a counter-balance during climbing. Some even rotate their tails in mid-air when jumping between branches as this helps control the 'flight' and reduces spin. Monkeys that spend most of their time on the ground have lost their long, sweeping tails. On the floor these would probably be a disadvantage, something to pick up dirt and be trodden on. Species such as baboons have unimpressive limp tails, while the macaques have just the merest stump left.

Superficially there is little difference between living monkeys and prosimians. Both groups have such diverse forms that, visually at least, the dividing line is well disguised. Monkeys tend to use their forelimbs as hands more than lemurs, but the real differences lie in behaviour. Monkeys are much more likely to play, explore and invent as an integral part of their everyday lives. They are nosy and noisy, argumentative and bright – all intellectual traits that are part of our own make-up.

From early monkey-like animals sprang the apes that share so many of their ancestors' characteristics. There are two families of living apes, excluding humans: the Hylobatidae or lesser apes, made up of nine gibbon species, and the Pongidae, made up of the four great apes. Together with man, the two families form the super-family of Hominoidea. Fossil evidence of ancient apes dates back as far as 20 million years ago. This means that modern apes did not evolve from modern monkeys but the two shared a common ancestor before they split and went their own evolutionary ways. The differences in local habitats may help explain the divergent paths taken by Old and New World primate species. The tropical belt in South America is heavily forested and appears to have been unchanged for millions of years. This would certainly have encouraged the appearance of specialized physical adaptations such as prehensile tails to fully exploit the environment, while the dense forest would have been less likely to produce heavier, earth-bound species along the lines of gorillas and chimpanzees. On the other side of the Atlantic in Africa, along with rain forest there are huge areas of grassland and broken woodland, favouring those animals which could compromise and take advantage of food and movement at ground level and in the trees.

Physically, apes differ from monkeys in that they have no tail; some monkeys do appear to be tail-less, but close examination always shows a remnant, no matter how small. Apes have forelimbs that are longer than their hindlimbs and their wrist joints are more flexible than monkeys', allowing greater hand mobility. Gibbons are the smallest of the apes; the siamang is the heaviest species and weighs in at a maximum of 11kg (24lb). Gibbons are confined to the forests of South-East Asia, where they spend their lives almost entirely in the trees. They illustrate perfectly the typical long-armed ape structure: their forelimbs are longer than the total length of hindlimbs and body. Their hands are so well suited to swinging through trees that they have effectively turned into living hooks. In true brachiation, swinging from branch to branch, thumbs are not only redundant but can also be a real nuisance, because they snag on twigs during the rhythmic hand-over-hand, grab-release-grab travelling technique employed by gibbons. As the lesser apes became more arboreal, their thumbs were almost abandoned. The gibbon thumb has now become so small and has moved away so far from the main four digits that it is almost totally ineffectual as a mechanism for manipulation. Gibbons can't use their hands in the same way as humans and the great apes; they lack both a power and a precision grip. When picking up small objects from the floor, a gibbon's hand acts as a soup ladle, scooping rather than precisely picking. Their tree-top speciality has cost them one of the skills that makes the larger apes so successful. Interestingly, gibbons do not score anything like as highly as the great apes in intelligence tests.

It would be far too simplistic to put this down to a non-functioning thumb, but it is a factor to be considered. Perhaps their unique anatomical

The gibbons of South-East Asia are classified as lesser apes. They have taken a specialized evolutionary path and are perfectly designed for a life in the high forest canopy.

development and ecological niche have resulted in a brain that has no need to be creative. Gibbons belong in the trees but are not helpless on the ground, and can even walk upright on their hindlegs for a short time. But they are not well suited to travel any distance like this and will quickly take to the trees if danger threatens. Gibbons are monogamous breeders and there is little weight distinction between the sexes, although their coat colour often differs. Most species are predominantly fruit eaters and the remainder of their diet is made up of leaves. The Kloss gibbon is a real exception: three-quarters of its food is taken as leaves and the rest consists of assorted animal proteins such as invertebrates. Although classified as lesser apes, the gibbons' small size means that they directly compete for territory and food with monkeys instead of their closer relative the orang-utan, which shares their habitat on Borneo and Sumatra. The territorial and courtship 'songs' of gibbons are some of the most memorable of all sounds that echo through Asian forests. Each species has its own call and when a pair or whole family group join in the chorus, the woodlands ring for up to half an hour. With their high-speed tree-top lifestyle, gibbons are not easy to study in the wild. We do know, however, that five of the nine species are seriously threatened and their future is gloomy if deforestation continues at its present rate.

We do not understand every turning and cul-de-sac in the progression from our earliest ancestors to modern apes and man; the evolutionary family tree can be drawn in several different ways and still give the same result. Evolution was once believed to be a linear process with 'lesser' species branching off and stagnating at an early stage. We now know it is a lot more convoluted than that. What we see in our own world is simply a thin cross-section of time; like cutting into a lump of cheese, only a tiny fraction of the whole is visible.

Current fossil evidence indicates the existence of three early primates: *Dryopithecus*, *Gigantopithecus* and *Ramapithecus*. These probably emerged from a group known as *Aegyptopithecus*, which appears to be one of the common ancestors shared by all living primates. *Gigantopithecus* was probably an evolutionary dead end, for they seem to have culminated in a group of huge orang-utan-like apes that eventually died out leaving no descendants. *Ramapithecus* was the first in the emerging line of early humans, while *Dryopithecus* (woodland ape) was possibly the emergence of an ape-like animal on the branch of development that would eventually lead to man. It is likely that *Dryopithecus*, once known as Proconsul, evolved in East Africa and from there spread into Asia and Europe. There is a mass of conjecture and disagreement about what happened at this stage. Some scientists are convinced that apes are all directly descended from *Dryopithecus* and man from *Ramapithecus*; others would stake their reputation on the impossibility of this.

Dryopithecus had a long thumb rather than the shortened version seen in modern apes, and, although we cannot be certain about the muscle

Orang-utan forehand.

arrangement, this would probably have worked with the same precision as our own. Although some scientists are convinced that *Dryopithecus* is a direct link in our evolutionary chain, others feel that the animals' scimitar-like teeth are so far removed from our own that there could not be a direct connection. The scarcity of usable fossil remains leaves us groping in the dark after this point. The divergence of the three genera is not disputed, but where each ended is not so clear-cut. The clues can be read in different ways and interpretation is a minefield. One fact is unchallenged, though: humans and the four ape species that we recognize today have come from a single shared ancestor.

Fossil remains give the impression that early apes were non-specialists, moving and feeding like modern orang-utans, which grip on to branches and hang from any convenient hand-hold. They were heavier than the prosimians and their extra weight would have forced a change in

With long hands and short thumbs, primates lack the precision grip and fine manipulation that is so important to man.

locomotion. Modern monkeys scurry and run over branches like squirrels because their body-weight is often very low. Apes, with their greater bulk, could not possibly maintain balance if they moved around a tree in the same way as they walked on the forest floor, so they have been obliged to adopt different techniques. Instead of keeping their bodies parallel to the branch like small tree-dwelling animals, they keep their backbone parallel to the tree trunk and rely on powerful limbs for support.

As with all family trees, the main trunk of the ape/human story can be either central to a simple diagram or a minor subplot that is just a small part of a much larger picture. At many points along the way, groups splintered off and explored their own evolutionary fate to arrive at the species we know today. Unfortunately, there is little fossil evidence covering the history of the African apes, so their particular strand of the tale has to be partially guesswork. However by using modern scientific techniques it is possible to gauge how closely species are related by analysing the make-up of their DNA. Deoxyribonucleic acid is the blueprint for all living matter. It is the starting point of life and is present in the chromosomes inherited from our parents. DNA dictates whether a newly fertilized egg will grow into a pig or an elephant; it dictates eye colour and even abstract features such as mating behaviour.

DNA is not immune to the effects of time and evolution; its structure changes just as the form of each species gradually alters. As apes, monkey and humans sprang from the same rootstock, it is possible to assess how closely related they are by looking at how much their respective DNA has altered. There is a yawning chasm between the DNA structure of a pig and that of a fish because they parted company, on the evolutionary tree, millions of years ago. But DNA analysis indicates that, in biochemical terms, humans and apes are remarkably similar.

As the structure of DNA appears to evolve at a remarkably constant rate, we can work out mathematically the time at which humans and apes branched off and make a fairly confident guess about the timetable of divergence of new species from the main evolutionary branch. Biochemical analysis suggests that the proteins of chimpanzees and humans separated around 5 million years ago. This gives us a fair idea of when the last common ancestor of both species walked the earth. From this point chimps and humans developed in their own ways. Gorillas branched out earlier, possibly 6 to 8 million years ago, orang-utans 7 to 10 million years and gibbons 12 million years ago. But be warned; these figures are not universally accepted.

Further analysis proves that chimpanzees and humans share 99 per cent of the proteins that make up their bodies; some, such as haemoglobin, are virtually indistinguishable. This is a higher comparative figure than the proteins shared by chimps and orang-utans, possibly suggesting that chimpanzees are more closely related to us than they are to their red Asian cousins. This apparently strong relationship supports the theory that humans did indeed evolve in the African landscape, and from this habitat came two surprisingly similar – and related – animals. Some zoologists believe that the family tree should be redrawn and chimps should be included on the same branch as humans.

Other researchers use different criteria which may be just as relevant. Orang-utans and humans share 41 per cent of parasites, while chimps and humans share just 33 per cent. Parasites are extremely sensitive to factors such as body chemistry; it is essential for survival that they choose the right host. This indicates to some primatologists that we are more closely related to the orang-utan than we are to the chimp.

In ape terms alone, there were many successful forms around before those we see now and, unless human stupidity wipes out the existing population, it is almost certain that from these species will come unknown apes of the future. Guessing at the nature of animals to come is a safe game, for the process is so slow that no one is likely to be around to witness the changes or challenge the predictions. Certainly there was a time when animals were in a race to get ever larger, dinosaurs being the supreme example. There is slight evidence that many species are beginning to decrease in size. The reasons for this could be totally artificial; selective hunting of the wrong sort certainly has this effect. For centuries the most sought-after trophies came from the biggest males; giant elephant tusks and tiger skins from the Victorian age show animals of a size rarely seen today. The largest were eagerly shot, but left alone it would have been these individuals that were most likely to breed and pass on 'big' genes to their offspring, ensuring the next generation would have been equally large. Consequently, hunting the most impressive specimens had the effect of weakening the whole population.

There is also the possibility that animals have stopped becoming larger because the area of their habitat has reduced. The endless forests have disappeared, space and food are at a premium and a large body-size could soon become a disadvantage. All of these factors are theoretical, but they have encouraged many zoologists to predict that apes of the future will be smaller than those of today. Several scientists have seriously put forward the idea of genetically engineering human beings to make them smaller by up to a third in order to minimize the problems of overpopulation. Each would then take up less room, use fewer natural resources and require smaller amounts of food.

The nature of evolution is still not understood. Some see it as an ever-changing process in which all species are constantly experimenting and mutating. Others believe that it takes place in sudden spurts, with long gaps of stability in between, until there comes a point when the species has nowhere left to go, either in physical or in evolutionary terms. It is possible that apes have now reached this stage, for there are no empty forests to colonize and man is encroaching rapidly on the little they have left.

2
THE CHIMPANZEE

CHIMPANZEES ARE ONE of the world's most instantly recognizable animals, even to those that don't have even a passing interest in wildlife. Their infiltration of twentieth-century popular culture has been universal and absolute. Chimps, or inanimate versions of them, feature in television advertisements, movies, jokes and toyshops all over the world. They are always in the top three attractions of every zoo that keeps them. It is not an exaggeration to say that human beings are quite simply fascinated by chimpanzees.

The name 'chimpanzee' is used to describe four different animals, all living in Africa. There are three distinct subspecies of the common chimp and nowhere do their ranges overlap. The western or black-faced chimpanzee, *Pan troglodytes troglodytes*, is found in Guinea, Sierra Leone and several other West African countries. It has the smallest population of the three. In Cameroon, Gabon, Congo and nearby states lives the pale-faced or masked chimpanzee, *P. t. verus*. Further east, in Zaire, Uganda and Tanzania, is the eastern long-haired chimpanzee, *P. t. schweinfurthi*, and this has the largest population. The physical differences between the three are well defined but of little real importance. Their long-winded names are probably enough to describe the features of each subspecies. Due to the infinite variety in the colouring of individual chimps, it would be possible to transport almost any animal into a different geographic zone, where it would probably go unnoticed by all but the real experts.

Most zoos have a motley collection of chimpanzees whose ancestors arrived through a chain of several dealers or via other institutions, and their family tree is usually unknown. Different subspecies, particularly in earlier days, were just lumped together as 'chimps'. Since then they have bred and interbred so often that any geographic race differences have vanished, leaving just hybrids in their wake. Only a few zoos have rigidly stuck to type and ensured that the subspecies remain distinct.

Historically there were up to fourteen 'recognized' species. This has now been whittled down, but even the three current subspecies are the

Young chimpanzees have pink faces that gradually darken with age.

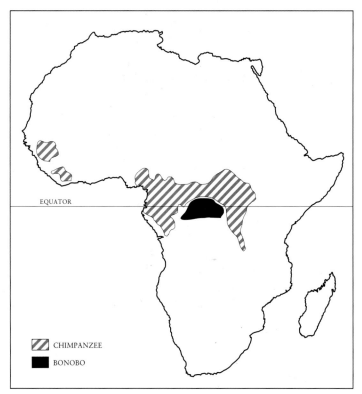

Fig 2
The distribution of chimpanzees and bonobos.

subject of debate. Convincing cases are made for a reduction or enlarge-
ment of the current list. Some researchers refer to a fourth race, the
koolo-kamba or gorilla-like chimpanzee (see Chapter 10), which occupies
the same area as the black-faced chimps. These creatures are reputed to
be jet black with huge feet and unusually thick hair. Koolo-kambas, as a
race, are not acknowledged by many zoologists; if mentioned at all, they
are classified as a variation of the resident black-faced.

Due to the nature of their habitat and scattered population, plus the
extreme difficulties of counting black animals living in the shadows of a
woodland, we do not know exactly how many chimpanzees now live in
the wild. Some estimates go up to 200,000; many are much lower. One
thing is definite. At the beginning of the twentieth century the figure
would have been around 5 million or more. Chimpanzees are becoming
more endangered by the minute.

Bonobos (*Pan paniscus*) are the fourth member of the genus. Until
recently these animals were known as pygmy chimpanzees. A recent
and welcome trend for tidying up zoological nomenclature resulted in the
dropping of the 'pygmy' adjective as it was very misleading. The
word chimpanzee is now rarely used for this species, with bonobo rapidly
taking over. Superficially, bonobos look very similar to common (a word
that is also becoming increasingly inappropriate as the population

dwindles) chimps. They are about the same height at the shoulder but their overall build is more slender and leggy, giving a body-weight that is up to 20 per cent lighter. They have much rounder heads and their muzzles are flatter; they also have smaller ears and teeth. Bonobos are born with black faces and hands, while those of common chimps are usually pink. Adult bonobos are blacker all over than other chimps, and they often have strange hairy tufts on either side of their head that are vaguely reminiscent of the remaining hair of a balding human.

The appellation 'pygmy' was given after an initial examination of the skull and bones rather than of a live animal. With a smaller head and more slender bones, the early evidence certainly pointed to a species that was simply a diminutive version of the common chimp. Despite similarities in both anatomy and behaviour, bonobos are as distantly – or as closely – related to chimpanzees as humans are. Despite this, bonobos were not seen as a distinct species until 1929, and this distinction was made only from skins and skeletons. They went unnoticed simply because they inhabit the dense and inhospitable lowland rain forests of Zaire, one of the last places in Africa to be scientifically explored.

Chimpanzees are among the most thoroughly studied animals on earth, for it has long been thought that by understanding these, our closest living relatives, we would better appreciate our own evolution and behaviour. Of the great apes, chimps are the nearest to us in size, social structure, adaptability and behaviour. But there is still some debate about whether or not they should be termed our 'closest relatives'. It is only in recent times that we have known anything about their life in the wild. The first true fieldwork was carried out by an enterprising, if misled, man called Garner. He built a strong cage in which he could safely sit and watch the everyday

A bonobo (right) has a slighter build and a smaller head than the more powerful common chimpanzee (left).

Of all the great apes bonobos are the closest to humans in appearance.

habits of wild chimps. For four months he patiently sat and waited, but the chimps were so frightened of the strange contraption that few would venture anywhere near. Those that did, scuttled past as quickly as possible and Garner learned absolutely nothing of their behaviour. Thanks to the less intimidating fly-on-the-wall approach of modern-day researchers such as Jane Goodall and Adriaan Kortlandt, we know much more about the chimpanzees' private life.

Chimps have muscular bodies with long arms and short, slightly bowed legs that give them a sloping back. Males are larger than females and tend to be around 30 per cent heavier (males weigh around 43kg/95lb, females 32kg/70lb). Apart from proportional development in teeth and muscles, along with obvious sexual characteristics, there are no major differences between the sexes. This is another factor that they share with humans, in sharp contrast to the size differences between male and female gorillas and orang-utans.

Chimpanzees occupy a wide spectrum of habitats, ranging from dense primary forest to open grassland, and almost everything in between. Their hearing and eyesight are about as well developed as those of an average human, although their sense of smell is slightly better. Chimps are quadrupeds that walk just like a gorilla, with hindfeet flat on the ground and forefeet resting on flesh-padded knuckles. They can stand and even walk upright on their back legs, but this is obviously not a comfortable position and produces a plodding, bow-legged gait that is kept up only for a short time.

Common chimpanzees have evolved as all-rounders, equally at home in the trees or on the ground. In trees, their long arms and powerful muscles give them adequate climbing skills but they are slow and ponderous. Chimpanzees lack the arboreal confidence and grace of the bonobos – for these animals are rain-forest specialists that are comfortable on the ground but, through necessity, spend more time in the trees than their heavier cousins. Bonobos venture high into the canopy in search of food and are just as likely to travel through the trees as they are on the forest floor. It is the need to leave the ground and climb to where most of the food is found that has brought about the bonobos' lighter and more agile frame.

Both species are highly sociable. Common chimps live in loose, extended groups that can include more than 100 animals; bonobos live in much smaller groups that often consist of less than twenty individuals. Social bonding in chimps is not as strict or structured as in gorilla families and the nature of their association is very complex. All animals within a population know the other members of the group but most day-to-day activities are either solitary or take place in small splinter parties that constantly form, disappear and re-form inside the larger group. Short-term groupings can last for anything from an hour to a week and this constant fluctuation of structure has been termed 'fusion-fission'. Dominant males are the stable core of a common chimpanzee group and they rarely move on, but other animals come and go unpredictably, and females particularly are liable to keep changing their social alliance.

Most chimpanzee groups have set ranges with well-defined borders but some have a migratory lifestyle, covering a much wider area in a wandering fashion that seems to have no real pattern. These groups may inhabit a region as large as 400 square km (150 square miles) stopping in one place just for a few days to exploit the food resources before moving on. Little is understood about the structure and life history of these migrants, or what effect they have upon the sedentary groups they encounter on their travels.

Bonobos are even less bound by social ties than common chimpanzees. And, unlike common chimps, it is the bonobo females that form the backbone of a group rather than the males. Their society is based on mutual co-operation rather than an enforced hierarchy maintained by powerful individuals.

It is always dangerous to draw similarities between human behaviour and that of other species, but sometimes comparisons are impossible to ignore. Both chimpanzees and bonobos form relationships that are exact parallels of our own. They have 'friends' with whom they prefer to eat and play; male-to-male bonds are the strongest in common chimps and close

A male chimpanzee in full display.

female bonds are rare. For some reason it is the exact opposite in bonobo societies. One possible explanation for this disparity might be that food is readily available in the bonobos' habitat. An entire group can feed freely from any one of the giant trees that pepper the forest, so there is little competition between individuals. Common chimpanzees' food is harder to find and is usually scattered over a wider area. Food has to be actively sought out and this creates a more competitive atmosphere; close bonding would be one way of reducing tension among the volatile males.

Individuals of both species have enemies within the group whose presence is enough to prompt an attack or a rapid withdrawal, depending on their status. They also show some of the less savoury aspects of human behaviour. Bullying frequently occurs when a small group of chimps, for no obvious reason, beats up another animal. This can grow beyond casual violence, with the gang making repeated attempts to catch their victim, chasing him through the bush for long periods. Should the unfortunate animal reappear, he will be attacked again and, if caught, the injuries leave no doubt that this is far more serious than a mere game.

Within the wider territory of each sedentary group, up to 50 square km (20 square miles), individual animals have their own core ranges that extend out to around 4 square km (1.5 square miles). About 80 per cent of each chimp's time is spent within their own inner territory. Chimpanzee groups are controlled by a powerful dominant male that has to constantly defend his position in the hierarchy against other high-ranking males. Uncontrolled male aggression can end in appalling injuries to all concerned and can even result in deaths. To reduce friction, males regularly reinforce friendship bonds through mutual grooming and group feeding. They are far more gregarious than females, particularly with their own brothers but also with non-related animals. They often stay together for many days in all-male bands of up to forty animals. These groups are transient and highly flexible but the close contact and mutual reassurance help cool down potentially dangerous situations that might develop between rivals.

Despite mechanisms to minimize violence, it frequently erupts. The first sign of male aggression appears as a rhythmic sideways swaying which gradually becomes faster and increasingly exaggerated. Eventually the chimp stands on its hindlegs while screaming at ear-splitting level with short, piercing shrieks. Body-hair stands on end and he rapidly slaps his body with open hands; he might also thump the ground and anything else that is nearby. This is followed by foot pounding on the floor in a fast drumming action. A chimp in a tree will stamp on a branch with his back legs and shake the foliage with his hands. The final stage is to hurl any convenient missile, usually sticks and stones, at the object of his rage. Chimps have admirable aim, with an uncanny knack of hitting their targets extremely hard.

Exhibitions such as these are not always brought about by real enemies. Males have been known to produce full displays when the weather suddenly takes a turn for the worse. It is tempting, and probably not too far from the truth, to see this as an ordinary tantrum. For some unknown reason, temper displays can be triggered by the sound of running water. It is certainly hard to imagine quite why a chimpanzee would throw a full tantrum at a waterfall, but there is no doubt that on occasions they do.

Dominance struggles between males often end at the display stage, even though both protagonists may keep up a dramatic performance for half an hour before one backs down. Submission is shown by a smoothing of the coat and dropping on to all fours again. Lips are pulled back in a wide 'grin' and a hand reaches tentatively forward to touch the victor. This is not just subservience; it's also a sign that the 'beaten' animal has a real need for reassurance. If the victor does not respond with a hug or friendly pat, the vanquished animal may follow him around begging and crawling with outstretched hands until some physical contact passes between them that heals the rift created by the original dispute. With soft panting grunts, the now subservient male will often groom his one-time enemy as a sign of friendship and subordination. Non-violent battles such as these ensure that the strongest animals stay at the top, while avoiding any injuries to the big males that would weaken the defences of the group as a whole.

There is a dominance ladder which takes in every male in the group, with the youngest and weakest at the bottom. The ladder is very fluid and individuals can move up and down freely, through changes brought about by maturity, ageing or injury. Unlike most other sociable primates, there is no way that the top male can be recognized physically. They are not obviously bigger or heavier, and they lack the 'colour badges' worn by silverback gorillas or dominant mandrills. Nor can high-ranking males easily be distinguished by behaviour. They achieve dominance not only through physical strength but also through intelligence and a subtle understanding of how to control group behaviour. A dominant male – known as the alpha – often has two or three prime males that support him. This is useful when there is a challenge for authority from a male immediately below him in the pecking order, known as the beta. He too will have supporters, and the stage is then set for a battle between two teams instead of two individuals. Cunning politics are constantly at work in the alpha and beta camps, both attempting to encourage the biggest males on to their team, while intimidating those on the opposing side. The possession of a powerful team of heavies can swing a finely balanced leadership battle one way or the other. Dominance has little bearing in everyday activities. One of the few advantages consists of the choice of good feeding sites. A high rank can bring more responsibilities than perks, for it is the big males' job to defend the group from attack and therefore they run a greater risk of being injured.

As chimpanzees may live for up to forty-five years, every male, barring genetic or accidental physical weaknesses which would keep him on a permanently low rung of the hierarchy ladder, has a very good chance of achieving high-level status at least for a short time. A prime male can expect to occupy his position for eight or ten years, but the alpha male is usually supreme boss for much less than this. As muscles get tired and adolescents get bigger, each male is eventually displaced in the pecking order. When once-dominant males lose their status, they are treated surprisingly well by those that have deposed them. In other primate societies, old males may be driven off, but chimpanzees are more tolerant. Ageing males are treated with respect, enjoying many of the privileges they had in their prime. This may be because they retain knowledge of feeding grounds and other details that are still of use to the group as a whole.

Like humans, elderly chimpanzees become wrinkled and bald.

Generally, adult females are submissive to all adult males of any status, but they do have a pecking order of their own. Displays may not be as exaggerated, but females can be almost as aggressive to their own sex as males are to theirs. Female status is a more subtle affair, sometimes pure aggression rather than physical size taking an individual up the hierarchy. The company of a large son or high-ranking consort will temporarily boost a female's position, for they are likely to defend her if a rival or more dominant female becomes too pushy. Low-status females will occasionally seek the protection of big males if they are attacked by others of their own gender. The males seldom take any notice, but their presence alone is enough to cool down the situation.

Unlike many other primate societies, it is not only the dominant male that gets the chance to mate. Chimpanzees, and particularly bonobos, use sex as a form of mutual bonding. Throughout the animal world copulation is a means to an end, and that end is reproduction. If the female is not receptive and cannot conceive, mating is a waste of energy. Chimps are different, for, like humans, they have adapted basic sexual behaviour and incorporated it into social life. Both the common chimpanzee and the bonobo use mating as a form of communication within the group. It can be used as a greeting, or at times of great excitement, such as following a dominance display or if a supply of favourite food has been discovered. This behaviour appears to be a method of diffusing tension. Because it includes all dominance levels in both male and female hierarchies, sex is a highly efficient way of bonding the whole population and minimizing aggression. The nature of their sexual activity is put into perspective by the cycle of breeding females. Having given birth, a female becomes receptive to males long before she is physically able to conceive. This indicates that copulation is a social lubricant in addition to being for breeding.

The frequency of copulation in bonobos is the highest of any ape, yet their reproductive success is no greater. Females will mate at any time during their cycle, including during menstrual bleeding and when pregnant. Some researchers have suggested that female bonobos use sex as a way of manipulating males when there is competition for food. Observations in zoos show that males with possession of a particularly tasty treat will not share out of choice. But when a sexually receptive female approaches and they mate, she will then be allowed to join in the feast. Similar, if less detailed, behaviour has been seen in the wild. It has also been noticed that lower-ranking females mate more frequently than high-status females. This could be a method of appeasing the group as a whole and therefore avoiding a situation where the animals at the bottom of the ladder bear the brunt of the accumulated anger. To return to the idea of sex as a social lubricant, this behaviour is very much like throwing oil over waters before they become troubled.

Both male and female bonobos indulge in sexually related behaviour with others of the same gender. Males engage in penis fencing as part of their play. This involves rubbing erect penises while hanging from a branch. The female equivalent is known as GG or genito-genital rubbing. This is a bonding ritual that is unique to bonobos and takes the form of two females sitting or lying face to face, with arms around each other while their genitals are rubbed together. The females move their hips just as they would when mating with a male, and occasionally one will produce a piercing shout that males normally give out during copulation. Both activities can be seen as group-bonding behaviour and therefore are just the same as the majority of heterosexual activity. It is possible that this constant mutual reassurance helps to maintain the bonobos' peaceful communal lifestyle.

Adult females come into oestrus about once every five weeks. When they become receptive, the skin around the genitals swells enormously and sticks out like a pink balloon full of water. This flamboyant advert attracts the attention of every neighbouring male. During the average ten-day oestrus period, a female will probably be mated by each adult male in the group, several dozen times each. Bonobos are receptive for far longer than the common chimpanzees. A female may retain her genital swelling for up to half of every cycle and therefore mate more times. One female has been recorded as being receptive for fifty consecutive days.

For most of the receptive period mating is completely promiscuous, but around the time of ovulation just a handful of dominant males compete directly for access rights; lower-ranking animals are swiftly threatened to keep them away. Dominant males need only to stare at a receptive female, or maybe throw a stick, and the implied threat is enough to persuade her to turn around and present her rear for mating. If she does not respond quickly, a male will enter a real threat display: branch shaking and hair erection will soon force the female into submission. Mating in bonobos is not quite as dramatic: intimidation displays are rare and at least a quarter of all copulations are instigated by the female. In both species copulation is brief, on average lasting seven seconds with common chimps and fifteen seconds with bonobos. Females may mate with five different partners in less than an hour.

Youngsters that are unable to take part in the proceedings show deep interest in mating pairs. They get very excited, scream, jump around and daringly attempt to part the couple. Offspring of the mated female are the most determined in their efforts to interfere. They will even start to desperately slap the male, a risk they would never normally take under more calm circumstances. This is not merely cussedness but has a real value: if a youngster can delay its mother's pregnancy, it will demand her individual attention and protection for longer. Males have a surprising tolerance for such disturbances and ignore most of the attempted interference, apart from actions from older juvenile males; these are treated as potential rivals for the female and may be chased off or attacked.

Male chimps have a mammoth capacity for mating. As they live in such

Receptive females advertise their readiness to mate with an unmistakable swelling of their rear end.

large groups, there are always receptive females available and a dominant male may mate several times a day for many years. They first begin to mate around the age of two years and will continue throughout their lives. To accommodate this high level of activity, males have proportionately larger testes than any of the apes (including humans) and an equally high production of semen. Mating within a chimpanzee group lacks any major importance; it is more a method of communication and mutual bonding between the sexes. Both male and female will continue to eat during copulation and once it is over, both will wander away as if nothing had happened.

Temporary but vitally important pair bonds are forged when a male and female choose to leave the main group and take to the bush. Bonding may last several weeks and, although they result in only around one-quarter of the mating activity, these periods of absence are responsible for most chimpanzee pregnancies. There seems to be no connection between a female's choice of a temporary consort and his position in the hierarchy of the group; low-ranking males are just as likely to form a bond as higher-ranking ones. With this free-for-all mating system, no male can be certain which are his own offspring. It is in his interest, then, to protect every infant in the group to ensure that his own genes are successfully passed on to the next generation. This system is very different to social organization in other group-living primates where one of the major advantages of dominance is the exclusive right to mate with all available females. Chimpanzee dominance has little bearing on their chance of fathering young; access to 'community' females at ovulation is of little value if most conceptions take place during pair-bonding trips. The random fashion of mating between all levels of chimpanzee society ensures that most males will breed successfully during their lifetime. The same cannot be said for the lower-order males of many other primate colonies.

Female chimps first come into oestrus at around the age of eight but they may not breed for a further two years. Zoo records indicate that captive animals breed up to three years earlier than their wild counterparts. Gestation in both common chimp and bonobo is around thirty-six weeks but there is considerable latitude in both directions with this figure. Single babies are the rule and twins are very rare. Due to the long dependence period of each youngster, a female will produce only one infant about every five years. This gap is reduced if a baby dies soon after birth. Newly born chimps are remarkably helpless and have only a weak grasping reflex; when travelling with the mother, then, they are in real danger of falling off unless supported by a maternal hand. A strong grip develops within a few days, allowing the baby to hang on to its mother's chest without assistance. At birth bonobos weigh around 1.2kg (2½lb) and common chimps around 1.5kg (3lb).

Unusually, the female often totally ignores the umbilical cord. Once delivery is over, the baby can be left attached and drags the placenta around until it dries and eventually drops off. Many other mammals chew through the cord and some even eat it as a rich source of iron. In survival terms, an attached placenta is potentially dangerous; the smell can attract predators and there is a risk of infection entering the baby's bloodstream. It is not known why some females eat the placenta while others ignore it completely. The long interval between births means that the arrival of a baby is not a common occurrence, even in a large group of chimps. The event attracts a lot of interest from all group members as a result. The short time following birth is one of the rare occasions when females will stand up to even the most high-ranking male. As he barges in to take a look at the newcomer, a female may either hold the baby close to her chest while keeping her back to the curious male or simply run away. The initial overprotection soon dies down, though, and each chimp then gets a chance to examine the baby, but the process is closely monitored by the mother and any rough handling is reprimanded with a short warning bark.

At times of stress or excitement even large offspring will hurry back to their mothers for reassurance.

For their first two years, babies suckle once an hour for two or three minutes, and they continue to take milk up to the age of about five. As they become older, more confident and physically stronger, infants take to riding on their mother's back instead of clinging to the underside. This appears to be 'fun', as most youngsters attempt to hitch a lift long after they are capable of keeping up with the rest of the group under their own steam. The desire to explore takes infants ever further away from the adult during rest periods and soon they join in games with other juveniles. But at the slightest hint of danger, real or imagined, they scamper back to the safety of mother. Sometimes infants are so absorbed by play and exploration that they fail to notice possible hazards, in which case the female briskly steps in and rescues her offspring. For the first year of their infant's life, females often stay close to adult males or form mother-and-baby groups. Both are strategies that offer protection from attack by a raiding chimp party or other predators; a lone female hampered with an infant cannot easily run or climb from danger.

At three years old, chimpanzees are infinitely more advanced than a human of the same age. They can climb and leap with confidence and will play-fight with any group member that comes close – and this includes big males that will put up with a barrage of poking, hair pulling and biting before walking away. Where territories overlap, juvenile common chimps have been seen playing with young baboons, while their respective parents maintain an aloof distance from each other. In these early tussles for food and power-play, it is the faster, more aggressive baboons that control every situation. Ironically, once the size and strength of chimpanzees outstrip those of their playmates, local baboons will be hunted, killed and eaten by their one-time associates.

At some time during the fourth year the young chimpanzee will experience a radical change in its mother's behaviour. She prevents it feeding by turning around or covering her chest with both arms. Any attempt to climb on her back will also be denied. This sends the young animal into paroxysms of rage. It throws tantrums, breaks branches and whimpers, but the female is generally unmoved. This is the time for the youngster to strike out on its own.

Noticeable gender differences in play behaviour appear very early on. Females show a deep interest in any infant and will spend long periods examining and grooming them. This fascination may be taken to the point where young females attempt to kidnap a baby and carry it around if its mother is not there to intervene. Males' play is made up more of mutual strength tests, such as wrestling and chasing, and they start to exhibit dominance displays from a very early age. They show, in miniature, hair erection and branch shaking during play, and males of less than two years old take great interest in receptive female adults. These will be closely followed and even mounted when the opportunity arises, and full mating movements are faithfully mimicked, but the chimps are not capable of full copulation at this stage. Young chimps begin taking solid food during their third year of life, and this is supplemented with milk for possibly another eighteen months before they are fully weaned. As chimpanzees give birth about once every five years, it is possible that a mother may have a new-born infant while still taking occasional care of a large juvenile. Interestingly, for the rest of their lives chimpanzees tend to spend more time with their mother than with any other member of the group.

To avoid constant inbreeding with close relatives, young female chimps usually leave their birth group and migrate to a neighbouring community. They may move two or three times before settling permanently. Females appear to test the water by 'dropping in' on a nearby group as a visitor. Mostly this happens when she is sexually receptive, to ensure that the resident males welcome her as a new partner instead of attacking her as an intruder. The females, however, see her in an entirely different light – to them the newcomer is a rival and they are quite likely to launch an assault unless the males intervene. Full transfer to another group may take several months of shuffling backwards and forwards. The departure of males would obviously have the effect of mixing genes, but it is not an option due to the aggression between males of rival groups. Adult males act as defenders of the whole community and they need to forge strong group bonds to ensure that they will fight to the death if necessary. A male that wandered from group to group would lack the motivation to risk injury on behalf of near strangers. So they stay put and the females move away.

Bonobos are less aggressive and violent than their cousins. Friction does occur, but it is over quickly with only minor injuries to show. They tend to use kicks and slaps in preference to bites, and even these are usually only superficial when they do occur. Inter-group conflict among common chimps is a totally different and unpredictable matter; the ranges of communities frequently overlap and the outcome of meetings is variable. All-male groups occasionally patrol the borders, looking for intruders. There is no doubt that this is the sole aim of the manoeuvre, for they sit at the edge of their home range and stare intently into enemy territory. They will walk the boundaries and make quick raids over the border before dashing back to home ground. When two groups meet, the event can be an anticlimax, all participants raising their hackles, screaming a lot, shaking branches and then backing off. At other times, though, bloody warfare can break out. When one group greatly out-numbers another, chimps will fight in small co-operative packs; two or three animals will hold down an enemy while the others bite and hit him. Staged ambushes such as this are one of the few real dangers to adult males. Armed with piercingly long canine teeth and flesh-crunching molars, a chimpanzee bite can inflict terrible damage on an opponent. On top of this they are fast, strong and extremely aggressive. They will attack enemies, of any species, with long branches wielded as clubs, together with missiles carefully selected for their weight. Experienced males go out of their way to collect large stones in preference to pebbles. These staged battles end with severe casualties on both sides and are probably essential to maintain the territorial rights of neighbouring groups, helping to ensure that trespassing is kept to a minimum.

The same lethal firepower is turned on to any carnivore that comes too close. There are a number of animals that might casually hunt an unguarded infant chimp; lions and crocodiles are both known to take unwary individuals should the opportunity arise, but only leopards will seriously attempt to hunt large chimpanzees on a regular basis. Being mainly nocturnal, leopards can approach under cover of darkness and pick off an outlying animal. Leopards seen in daylight do not escape so easily; all available chimps scream and throw stones, while big males move in close and attempt to hit the cat with fallen branches. The onslaught of missiles and noise soon drives off any leopard in a matter of seconds. Large snakes provoke a similar response, to the point where chimpanzees will even attack ropes and hosepipes that appear suddenly inside their territory.

Strangely, humans – the only other real enemy – are treated in a completely different way. The first animals to see the intruders give out short, barking alarm calls that instantly freeze the movement of the whole group. They keep completely motionless until the danger has passed. If seen, chimps in the trees drop to the ground, where they can move faster and melt silently into the bush. The desire to escape has probably evolved because in many areas chimpanzee meat is a prized delicacy. Over count-less generations they have learned that leopards can be driven off but humans intent on hunting are far more persistent and not as susceptible to a nerve-breaking display. Outright attacks on humans are rare, but they do happen. In direct conflict even a fully grown man would have little natural defence against a mature male chimp. Fortunately, most physical encounters are fleeting and involve a passing warning slap which may be painful but is far from lethal.

Common chimpanzees appear to really detest water in any form; they will travel miles out of their way rather than cross even the smallest river, and heavy rain tends to bring on what can best be called a slothful depression. During relentless tropical downpours they sit huddled beneath whatever shelter they can find, no matter how ineffectual. Most adopt a dejected squatting position, with hands held folded in front of their chest and head bowed. The chimps' long hair helps guide the water quickly downwards and each animal sits dripping and dreary for the duration. All play and feeding stop until the rain passes and then it seems as if the animals are genuinely elated by a return to decent weather. Unlike the other apes, bonobos show a human-like interest in water. Although they will not attempt to swim, in hot weather they will sit in shallow water or wade around looking for aquatic insects and plants to eat; they will even catch the odd fish by scooping it out of the water with a cupped hand. The fisherman then has to scramble on to the bank to retrieve it before the titbit is snapped up by another bonobo.

The chimpanzees' day begins shortly after dawn. They feed extensively for several hours until around noon, when the whole group stops for a sleep. The nature of this rest period depends on the time of the year and the gender of the animal. During the dry season many chimps just curl up on the ground; some females build platform nests in the trees whatever the weather conditions. When the rains come, the whole group will build day nests in the trees to avoid lying on the sodden ground. Although chimpanzees are often portrayed as noisy, quarrelsome creatures, there is a gentle and calm side to their personality.

Quiet midday rest periods may last for two or three hours and some of this time is spent in mutual grooming and playing. It is not too anthropomorphic to say that chimpanzees enjoy close personal contact as much as humans do. In moments of stress, no matter how old they are, most animals will run back to their mother for a quick hug. If she is dead or unavailable, contact with any nearby member of the group will reassure a frightened animal. Big males will even run to juveniles when they are badly shaken. This is not a call for help but a need to bolster their confidence by reminding themselves that they are not alone in their time of trouble.

This need for contact does not surface only when danger threatens; during rest periods mutual grooming takes over an hour. These sessions are longer than necessary just to keep a sparsely haired coat clean. Grooming is prolonged because the animals enjoy the process and like the attention. Hugging and patting are used when chimpanzees meet after an absence or when they become excited about a find of rare and tasty food. Touch is an integral part of their communication system and sometimes this turns into true play. All chimpanzees, when they are relaxed, enjoy being tickled. The stimulation of any part of their body might result in what can only be described as chimp 'laughter'. This emerges as short gasps made with a half-open mouth, produced while the animal squirms with pleasure. Tickling can occur between any animals, regardless of gender or status. And, just like humans, some chimps are more sensitive to thigh tickling or neck tickling; this is a striking resemblance to our own responses under similar circumstances.

Chimps use a wide range of facial expressions to convey emotions and these can be seen clearly in both wild and zoo populations. Extreme fear shows itself as an extraordinarily wide grin with lip corners pulled back to their maximum and jaws partly open. The fear-grin can be repeated many times over a short period, the frequency of its use a sure way of gauging the degree of the chimp's fright. Aggression is signalled with tightly compressed lips in an expression that is very similar to that used by angry humans. An open, pouting mouth, with lips pursed forward and

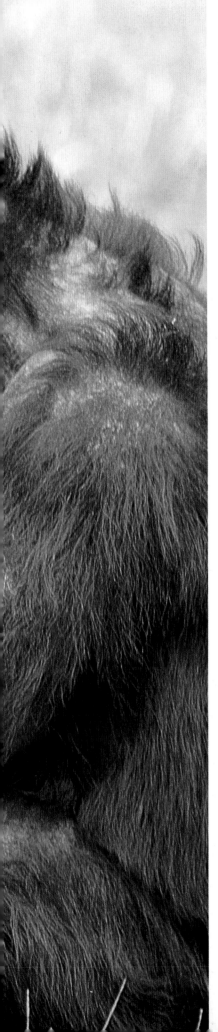

outward, is a sign of rejection or frustration. It is used when an animal is denied food or attention, and can be accompanied by a series of soft, plaintive hoots. Play faces are more difficult to pinpoint; they start as a half-open mouth with bottom teeth exposed and top teeth covered. If the game gets boisterous, the upper lips are pulled back to show the top teeth. But never are the teeth and gums exposed to the degree they are in a fear-grin.

Day nests are either flimsy or non-existent, but those used for the serious job of sleeping at night are more carefully constructed. The height of the nest is very variable, depending on local conditions, but most are high up in trees that are awkward to climb, as the position is the main defence against nocturnal predators. Nests can be built in the fork of a tree or on a network of closely growing horizontal branches. Using this as a base, nearby branches are bent inwards and are usually partially broken to prevent them springing back. Awkward branches are held down by the chimp's feet while it adds more undergrowth to the pile to keep them in position. If there is not enough vegetation nearby, branches and twigs will be collected from further afield and added to the bed. Some chimps are careful to remove any large branches that overhang the nest; this may be a safety precaution to ensure there is no easy route for predators to strike while the ape sleeps.

Some chimps show a high degree of weaving skill in nest building and this helps keep the structure intact for a long time; others merely pile the vegetation roughly together. Mature females appear to be the most accomplished nest builders: some are able to make a complex platform in less than five minutes. It is not unknown for males to sit patiently as a female builds a nest and then, once it is finished, commandeer it with screams and wild displays. The evicted female must then start again while the male drops off to sleep in her original bed.

Most chimpanzees are active until dusk and then take to their nests just as the last of the sun's rays disappear. They are up again with the dawn, unless it is raining, when they show a marked reluctance to stir at all. Sleeping chimps, like humans, adopt a variety of positions, but mostly they curl up on their sides. Nests are kept scrupulously clean. Even at night urine and faeces are carefully deposited over the edge, to avoid soiling the bedding material. Travelling chimps build a new nest each night, but when they remain in one spot a nest could be used for several days and repaired as necessary. Young chimps begin making nests at less than a year old, but these early attempts are very clumsy. Until the age of four or five they share their mother's bed, before taking the plunge and moving into one of their own. For some time after, many youngsters build their nests very close to that of their mother, probably for reassurance while they are finding their way in the world, alone for the first time.

Mutual grooming between adults takes up a significant part of every day.

3
THE ORANG-UTAN

A<small>T A RECENT</small> international conference devoted entirely to orang-utans, the species was described as 'the neglected ape', the implication being that gorillas and chimpanzees receive more attention, both politically and financially, than their Asian cousins. The red, shaggy orang-utan is definitely out on a limb in every way when compared to the other great apes, at least at first glance. They are the only non-African species, their appearance bears little resemblance to that of their close relatives and they lead an elusive, solitary lifestyle that is unique among the apes. Chimpanzees and gorillas have always been featured, both at popular and scientific levels, far more than orang-utans, probably because their sociability results in a lifestyle and behaviour that are much more like our own, helping us relate to them. This in turn has led scientists to think that chimps and gorillas offer more of an insight into our own behaviour. (There is also the more pragmatic reason that the ground-level activities of chimps and gorillas make them easier to study than orang-utans.)

Orang-utans are the largest tree-living animals in the world. Their range once covered much of South-East Asia, but now they are confined to the islands of Sumatra and southern Borneo (Kalimantan) in Indonesia, and northern Borneo (Sabah) in Malaysia. Current estimates give a world total of around 25,000 wild orang-utans with nearly 1,000 more in zoos and possibly the same number again being held illegally as pets or for unsavoury commercial purposes.

The two island races of orang-utans are separated by both distance and time. Close analysis of DNA suggests that the populations were split around 1.5 million years ago. Conventionally they are now regarded as subspecies: the Bornean race is *Pongo pygmaeus pygmaeus* and the Sumatran *P. p. abelii*. Not all researchers subscribe to this theory, however some arguing that the two are in fact distinct species. Chromosomal evidence indicates that the island races are as different from each other as chimpanzees are from bonobos. There is even a suggestion that orang-utans from western and eastern Borneo should be given distinct subspecies

A male orang-utan from Borneo showing the concave cheek flanges that distinguish it from a Sumatran male.

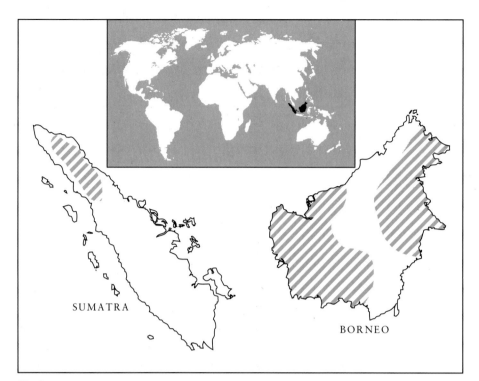

Fig 3
The distribution of orang-utans.

status. As can be gathered, taxonomy, the classification of animals into groups, families and species, is not a cut-and-dried business. In fact, it is usually in a state of flux, with diametrically opposed exponents arguing their case, which are often based on different criteria or varying interpretation of known facts.

Physically the visual differences between the two island races appear to be superficial, and individual variations confuse the issue still further. Sumatran orang-utans are lighter in colour and have thicker, longer hair than those from Borneo. This difference in hairiness could be misleading as a pointer to species identification, as the Sumatran animals live at higher – and therefore colder – altitudes, so their thick coat may simply be a local adaptation to compensate for the cooler conditions. An orang-utan from Borneo may well develop a thicker coat if transported to a colder Sumatran habitat. A by-product of this extra hairiness might be the reason for the noticeably shaggier face of the Sumatran male. These often have luxuriant, rich beards and 'moustaches' that extend out on to their cheeks. The beards of their Bornean counterparts are positively puny and underdeveloped in comparison; even Sumatran females have impressive chin hair.

The one outstanding feature of a male orang-utan's face, the cheek flange, is very different between the two islands. Cheek flanges are the fleshy discs that appear on either side of a male's face once he reaches maturity. They serve no practical function, but they give each male his highly distinctive huge-headed appearance. The flanges are simply a badge of rank and seniority, like the silverback on a gorilla. In Bornean orang-utans the flanges start to appear when the male is about eight years old and they are fully formed by the time he reaches fifteen. They grow outwards and forwards from the skull, giving a concave shape to the face. By comparison, Sumatran males are late developers; their cheeks start to expand at around ten years old and do not stop until the animal reaches twenty. Sumatran males have flanges that grow sideways rather than forwards, giving them a massive flat-faced look that bears little resemblance to the dish-faced Bornean apes. In both races males may remain subadult for several years beyond the point when they would normally reach a higher status; then, for no known reason, they go through a growth spurt, develop cheek flanges and move up on the hierarchical ladder.

The orang-utans' environment is made up of primary tropical rain forest where, on the ground, light levels are cut to a minimum by the dense canopy vegetation above. In a constant battle to reach the energy-giving sunlight, mature rain-forest trees can be astonishingly tall, with few branches on the lower part of their trunks. Leaves would serve no purpose in the lower, darker parts of the forest, so the best feeding, travelling and sleeping grounds are high in the trees, where most of the fruit and leaves grow. As orang-utans spend most of their time in trees, it should come as no surprise that they are extremely difficult to find, let alone study in detail. Chimpanzees and gorillas live in family groups that are active, noisy and relatively easy to find, as they leave a clearly visible trail. Both of these species also spend a high percentage of their time on the ground, making observation at least possible for human researchers. Orang-utans, on the other hand, are loners that make little noise as they slowly and confidently travel around high above the forest floor, out of sight and difficult to track. It was not until the 1970s that we really started to discover anything about the lives of wild orang-utans and nearly three decades later, there is still a lot more to learn.

Unlike gorillas and chimps, which crept into Western consciousness over many centuries through obscure classical references and jumbled eyewitness reports from early travellers, orang-utans were completely unknown to science until the eighteenth century. The first live orang-utan arrived in Europe in 1776, causing quite a sensation as it was so different from any other animal ever seen before. The hairy red ape was a great hit and there was a widespread call for more living specimens to be brought over. This task was greatly complicated by the language of the period, with both the orang-utan and the chimp being given the sweeping name of 'pongo'. Avaricious traders did not seem to worry that one animal was huge and red while the other was small and black. European collectors, placing their order from home, may have been specific in their instructions but they could never guarantee which species would actually arrive.

Chimpanzees were the most abundant ape normally offered for sale as more traders were operating in Africa than in Asia. And chimpanzees had a better chance of surviving the sea journey simply because West Africa is closer to Europe and they spent less time in small and unhygienic cages. Orang-utans taken from the wild had to withstand a brutal and tortuously slow sea voyage from the other side of the world, often left exposed on the deck for the whole passage. Most did not survive.

Linnaeus, the eighteenth-century scientist who was the first to classify animals into modern groups and families, originally believed that chimps and orang-utans were both members of the same species. He gave them the collective name of *Simia satyrus* - or wild ape man. Not surprisingly, a later researcher, J. F. Gmelin, realized that they were in fact two totally different animals. In 1788 the orang-utan was fully recognized by Western science.

Now, over 200 years later, there may be new information that changes our mind about the identification of orang-utan species. Recently there have been an increasing number of reports from central Sumatra that tell of a previously unidentified ape. This has been provisionally given the name orang-pendek or short man. Sightings have been recorded for over a century but evidence is scarce and confined to just a few eye-witness accounts. However these are sufficiently interesting to encourage further study by experts. It is not impossible that a large ape remains undiscovered in the dense forests of Sumatra.

Obviously, local people knew about the orang-utans long before Europeans set foot on the islands of South-East Asia. The name 'orang-utan' is usually quoted as a direct translation from Malay, meaning 'man of the woods'. It is regarded as impolite, particularly in Indonesia and Malaysia, to abbreviate the animal's name to 'orang', as this simply means person. There are many local stories that explain away the origins and presence of these uncannily human creatures. One tells of a man long ago who owed money to a neighbour, couldn't repay it and disappeared into the forest to avoid facing his responsibilities. With time he, and his offspring, became less human and more like the animals that shared their forest home. Stories such as these illustrate the esteem in which the red apes are held by the humans living alongside them.

There is no one single human culture that covers the same geographical area as the orang-utans, and each distinct group of people had their own view of the wildlife that surrounded them. Generally orang-utans were respected; many tribes believed that there was some physical or spiritual connection between humans and the red apes. Orang-utans were occasionally kept as pets in the communal long houses that were the centre of all village life. As youngsters they are playful and affectionate, but as they get older humanized orang-utans become increasingly unpredictable. Adult males are particularly changeable in their behaviour when they have been in captivity for long periods. He may look sluggish and docile, but a big male is capable of doing terrible damage should he chose. He can shift from a friendly acquiescent pet to an unstoppable avalanche of violence in just a few seconds. Orang-utans do not go through the screamingly violent temper bouts that chimps display. Their 'explosions' are far less dramatic and noisy, but they are every bit as dangerous. And often the change is not brought about by anything we can comprehend. Baby orang-utans are surely among the world's most appealing creatures and many people have been charmed into the idea of keeping one as a pet. They will all undoubtedly regret the decision in years to come, for close contact with an angry adult male is an experience that should be avoided at all costs. Orang-utans, like all the other apes, do not make good house-guests.

The Dayak people of Borneo call the ape a 'mias' and are well aware of the animal's potential strength. One Dayak hunter told an early European traveller: 'No animal is strong enough to hurt the mias, and the only creature he ever fights with is the crocodile . . . He always kills the crocodile with main strength, standing upon it, and pulling open its jaws and ripping up its throat.' This accurate assessment of the orang-utans' power never stopped the Dayaks hunting them for food, although with limited weapons they probably killed relatively few and had little effect on the population. Even today there are groups in central Borneo who still occasionally eat orang-utan meat.

Everything about the life of an orang-utan appears to run at an unnaturally leisurely pace. They move and eat slowly, produce only a small number of young in a lifetime and their lifespan is considerable, reaching an age of thirty-five in the wild and up to fifty in a zoo. Their apparent lethargy has always tempted first-time observers to judge that this is a reflection of a feeble brain. One early report tells the reader that orang-utans are 'as dull and slothful as can well be conceived'. A nineteenth-century children's book describes the ape in even less complimentary terms: 'The face of the animal is very ugly, like that of a very wicked old man, so that you can hardly look on it without dislike.' Drawing conclusions about an animal's intellectual prowess simply by looking at its appearance now seems ridiculous; it is a little like looking at a book and claiming it is dull because it's green. Those that know orang-utans well see beyond the bumbling external image and are fully aware of this ape's remarkable mental abilities

The male orang-utan, at about 90kg (200lb) may be almost twice as heavy as the female, who tends to weigh in at around 48kg (106lb). The size disparity, when added to other physical differences between the sexes, is so great that early explorers thought the males and females belonged to entirely different species. As the most arboreal of all the great apes, orang-utans have a perfect anatomy for life in the trees. Their body-weight is evenly shared between all four limbs; this is a useful aid for balance and helps spread the load on a tree. Shoulder and hip joints are remarkably supple. When walking on the ground an orang-utan uses its legs just as a dog does, but when sitting down the ape's legs can stick straight out sideways from its body, a position that few humans could hope to copy. This remarkable flexibility is clearly visible when an orang-utan climbs. Any of its limbs can be used in any position. I have even seen them eat while hanging upside-down.

In several areas orang-utans share the habitat with their distant, smaller cousins the gibbons. These really are the masters of high-speed tree-top travelling. With impossibly long arms, they swing at breathtaking speed from branch to branch, often leaping huge distances. Orang-utans are much more sedate; they have little choice with their much greater bulk. An orang-utan's long forearms are much better suited for reaching out to grip distant branches and then hauling their bodies behind. All four limbs are equipped with powerful grappling irons in the form of huge hands and feet that operate totally independently of each other, any one of which can support the entire weight of the ape.

When moving from tree to tree, orang-utans often use their weight to help cross gaps in the canopy. They hold on to a branch and start to swing slowly, building up momentum until their body is acting just like a pendulum. Then the moment comes when the swings take them far enough to grab another branch and move on. Alternatively, they climb as high as possible and then, again by using their weight, start to swing the entire tree until it bends far enough for them to catch hold of the next branch *en route*. Females with young frequently bridge gaps in the canopy with their own body to make climbing safer for the smaller and less confident infants. They hold on to two trees simultaneously while the youngsters clamber over them to reach the safety of the next big branch. Being much smaller than males, female orang-utans are more vulnerable to attack by large predators such as tigers. These cats are almost totally terrestrial, so female orang-utans stay out of their way by remaining high in the tree-tops. They will be caring for a youngster of some age for most of their adult lives and infants both slow them down and offer a tempting target for a tiger. Even high in the canopy orang-utans are not totally safe, for up here is the domain of the silent and deadly clouded leopard. This elusive cat is 1.5m (5ft) long and hunts mammals high in the trees. It has the power and speed to bring down a fully grown female orang-utan, and a preference for nocturnal hunting gives the clouded leopard a considerable advantage over sleeping apes.

The proportion of time orang-utans spend on the ground depends to a large extent on the size of the local tiger population. In Sumatra, where tiger numbers are at their highest, female orang-utans spend up to 95 per cent of their lives in the trees. Mature males, on the other hand, are adversaries to be respected by all carnivores. A big tiger is capable of killing a male orang-utan, but only in exceptional circumstances. Most would prefer to tackle less substantial prey and avoid the risk of being badly injured themselves. As with gorillas, the body-weight of male orang-utans

The incredibly flexible hip-joints of an orang-utan enable it to perform acrobatics that would be impossible for a human to imitate.

Young orang-utans climb into the high canopy to escape the attentions of predators. Despite their weight, these apes are surprisingly agile and balanced.

is a great disadvantage when climbing trees. This prevents them exploiting the high canopy, with its dangerously thin branches, and effectively reduces direct competition for food between males and females.

Even when travelling long distances females warily stay at canopy level where progress may be slow but at least it is safe. They go to great lengths to avoid setting foot on the ground, often taking complicated and circuitous aerial routes to reach a destination that could be approached far quicker at ground level. The need for females to escape into the trees is just one of the possible explanations for the enormous size difference in the sexes. Evolution has obviously decided that, for a female with young, a light, agile body is a better defence against predators than a heavy bulk. That could be because an animal the size of an adult male would be slow and may have difficulty protecting a curious and active youngster on the ground. Another excellent reason is that the really dangerous killers, tigers, can't reach the tree-tops.

Males may also be bigger to avoid direct competition for food with their females. As it stands, the sexes feed at different levels of a tree and only occasionally vie for the same food. According to the accepted theory of sexual selection, male orang-utans have inevitably grown larger than their females because there is regular competition for mates. Big males can hold territories against rivals; their displays are also more impressive, so they attract more mates and therefore large males have a greater chance of passing on their genes to successive generations.

Fully grown males are far more likely to clamber down to the ground, where their weight is less restrictive and they can move much faster. As we have seen, the main growth of primary tropical rain forest is at canopy level; little light filters down to the ground. The forest floor is made up of rotting wood and decaying leaves, and there are surprisingly few plants to hinder travellers of any species. Unlike other apes, orang-utans do not

Females with young spend most of their time high in the tree-tops. They venture on to the ground only to cross open areas or, near rehabilitation camps, to be fed by tourists.

look quite comfortable on the ground. They may be quadrupedal but they lack the thick fleshy pads that protect the knuckles of their cousins while walking on all fours.

Borneo has no resident tigers, so here orang-utans of both sexes are more likely to venture on to the ground. Also the Sumatran forest canopy is thicker, with a more interconnecting branch system making tree-top travel easier; in some of the undisturbed National Parks, a female orang-utan could travel many miles without ever once leaving the safety of the trees. But because feeding takes up so much of their daylight hours, few travel far in any twenty-four-hour period. Orang-utans move slowly and noisily around the trees and this makes inexperienced viewers think that the apes should be easy to find. In a zoo their bright, shaggy coat looks as if it should announce the apes' presence to all around. But in a dappled and dark forest the untidy hair melts in perfectly with the background and the orange colour merely looks like part of a tree and the curious epiphytic plants that grow on it.

Like other apes, orang-utans are accomplished nest builders. Day nests are usually hastily put together for a short siesta, while night nests are more robust structures. The choice of nest site depends upon the size of the builder: females and young males move high up into the canopy, making nests that can be up to 30m (90ft) above the ground; males are forced to sleep lower down, where the trees can bear their weight.

Orang-utans certainly have favourite areas for sleeping. They often pick trees on west-facing slopes so that they can absorb the last rays of warmth from the setting sun. They also like building nests on trees hanging over water, probably because they offer a degree of protection from nocturnal predators that may be tempted to climb up from the forest floor. Strangely, night nests are rarely found in food trees, although they might be in the tree next door. Day nests are sometimes built in the tree where an orang-utan is feeding, but at dusk they move out before settling. Most individuals build two new nests every day; old ones may occasionally be re-used but only after a new leafy lining has been installed. The nest itself is made up of thick branches piled around the fork of a branch to form a platform. Suitably well-leafed branches are torn from the tree as building material for the main structure, and for extra comfort a further layer of soft foliage is added on top. Despite their huge diameter of around 1.5m (5ft), new night nests are so accidentally camouflaged with green leaves that they are easy to overlook, although as the leaves turn brown the nest becomes more obvious in the green canopy. Newly independent youngsters are known to take the easy way out by sleeping in an old nest, and big males often do without completely, preferring to spend the night propped up in the fork of a tree.

Breeding behaviour in the other great apes is influenced by the fact that they live in large groups, but this is not the case with the solitary orang-utan. Apart from the early relationship between mother and dependent young, orang-utans spend the majority of their lives completely alone. Females reach sexual maturity at around ten years old, although the onset of a menstrual cycle may begin up to three years earlier. Males are capable of mating from about the age of seven but rarely have the chance to perform. Captive males have been known to sire young at the tender age of eight years old, but this has to be seen as unusual because in the wild a male of this age would be unlikely to attract a female in the face of competition from much bigger adults.

Females have a menstrual cycle of around twenty-eight days, and they are receptive to males at any time. The oestrus period, which is the time around ovulation when they are most likely to conceive, lasts around six days. As orang-utans are solitary, there is no point in the female visually advertising her readiness to mate. Female gorillas and chimps both show genital swelling around the time of ovulation and this is an open signal to the surrounding males that she is receptive. Such a sign would be wasted in orang-utan society, because usually there are no males close enough to interpret it.

Adult males have large territories that frequently overlap those of rivals, and each territory includes the home range of up to four females. Violent disputes among males over territorial rights are often avoided through the use of the 'long call', which is unique to orang-utans. This is a series of loud roars that are amplified by the male's huge throat pouch, which acts as a resonating chamber and exaggerates the sound. The result is a booming noise that can be heard for a good distance around. The sound level is similar to that of a lion roaring. Calls appear in the form of a series of short roars that get progressively louder before finally trailing off in a string of quiet groans. The whole sequence lasts less than two minutes. Males in Borneo have much larger throat pouches and their calls are drawn out; Sumatran males have pouches that are noticeably smaller and they produce a series of calls that are both shorter and faster. It is possible that the environment of each has promoted a difference in the sounds. Sumatra has a much more mountainous landscape than Borneo and we know that a series of short calls carries further through hills than a single long one. But in both cases the calls are used to re-inforce ownership of a territory and each male performs the long call ceremony almost every day to keep rivals aware of his presence. This non-visual signalling separates adult males and advertises their presence to any females within earshot. There is still a lot to learn about this last point, for some researchers have shown that females go out of their way to avoid an area containing a calling male. Male response to a long call varies according to the status of the listener. Subadults or males trespassing on to a rival's territory are likely to beat a quiet retreat if the resident orang-utan calls. A fully grown male within his territory will call back and often move towards the sound if he hears another male. Females with very young infants will climb to the top of the canopy to avoid contact with a calling male, while sexually receptive females will be drawn towards him.

If the quality of the calls is well matched and two adult males meet while trying to seduce a female, they will often fight over the right to mate. Both can be badly injured. Battling orang-utans use their feet and hands in combat. Bites are common and, with their flat, powerful teeth, the wound is a mixture of crushed and badly bruised flesh. Well-matched rivals also indulge in hand-to-hand warfare, where the two antagonists test each other's strength through a form of arm bending. Bones are regularly broken during these skirmishes and many adult males have permanently stiff fingers where fractures have healed badly. Facial scars, particularly on the cheek flanges, are another reminder of long-forgotten status battles.

Another display consists of breaking and dropping heavy branches. This acts as an audible clue to the physical size and power of the male. Big adults can smash the largest branches and the noise they make testifies to their strength. Although branch breaking generally takes place after a long call performance, it can also be used as self-defence. While photographing a subadult male in Borneo, I was showered with a barrage of sticks, plus one or two hefty branches. I was in no mortal danger from these missiles because the irate orang-utan made so much noise tearing them off that I had plenty of time to move. But some would certainly have been big enough to make me, or any other intruder, beat a hasty retreat if they had struck home. When really annoyed, orang-utans produce a sharp, loud 'sucking kiss', which is a warning to the listener not to take further liberties.

Branches are not the only threat from above. Earlier that day I had watched the same male as he dozed, draped untidily over the fork of a branch. I settled down for a snack from my rucksack and was happily tucking into a flapjack when the sleeping ape urinated copiously over my head and shoulders from 20m (65ft) up in the canopy. For the first few seconds I thought it was the start of just another sharp tropical downpour, but the distinctive smell soon dispelled that idea. This is an aspect of wildlife watching that is never portrayed on TV programmes, and was yet another reminder that working with animals is not the glamorous occupation it seems from the outside. If there is not a local saying that tells unwary travellers of this danger, then there really ought to be: 'It is a wise man who does not shelter beneath a tree in which sits an orang-utan.'

Courtship and mating behaviour in orang-utans is impossible to simplify or generalize about, for their sex life is almost as unpredictable and inventive as that of humans. Early mating research took place in zoos and this led to the popular belief that orang-utan males indulged in rape. With their greater size and strength, males would have little difficulty forcibly mating with a female, but the conditions in a cage are very different from those experienced by wild orang-utans. There is no doubt that males do use force in captivity where both animals are confined in a small area, allowing the female no escape route. Their rain-forest counterparts, being lighter and more agile than adult males, have few such problems. They just need to climb higher up into the canopy where no big male can follow. Rape may be an occasional option for an adult male in a forest, but it is far rarer than zoo observation would suggest.

This is not the end of the story. Males are eager to mate long before they reach full body-size. Subadults are almost as light and athletic as fully grown females, but they are much stronger. These young males can climb anywhere a female can and when she is receptive and a subadult becomes interested enough to follow, forcible mating is a likely result. Birute Galdikas, a Canadian researcher who has worked with orang-utans for many years, believes that this behaviour has little to do with reproduction; rather, she thinks it is just part of the learning process. Forcible mating by subadults does not seem to lead to pregnancy. It may simply be a mechanism for inexperienced males to rehearse the roles they will play in later life. Young males often approach a female in a gentle way and only when she is singularly unimpressed by their lack of size and status do things become more violent.

In a dense rain forest it would be possible for a female to elude a large, lusty male for months on end if she really wanted to. This would obviously be a disaster for the future of the species, so common sense dictates that females must voluntarily accept male advances at some stage. Receptive females sometimes elect to enter the territory of a calling male, and their choice of partner is influenced by his size and status. All of the information they need is carried in the sounds of the long calls and from these potential partners can be assessed from afar. This much is obvious, because large, powerful males father more offspring than small, low-ranking males.

Co-operative mating at its simplest is over in less than an hour. The pair meet, courtship is minimal before copulation and the two animals then part and go their separate ways. On the other hand, the courtship period can be stretched to fill many weeks. Mating itself can last up to fifteen minutes and may be repeated many times during courtship. Copulation is an infinitely varied feat, taking place anywhere in the trees or occasionally on the ground. The pair can be sitting in a nest, on a branch or hanging upside-down from a broken tree trunk. Male or female can be on top and the act is slower and less urgent than in any of the other species of ape. I hesitate to add this last point, for I can almost hear the shouts of 'anthropomorphism', but it really is impossible to watch mating orang-utans without feeling that they appear to have more sexual fun than their immediate relatives. The act can be initiated by either partner and the variety of positions is endlessly inventive, with either male or female taking the active role. The duration is astonishing when compared to the lightning acts of gorillas and chimps.

Rivers are physical barriers that prevent orang-utans moving freely around lowland rain forests.

On the two islands there seem to be major differences in the males' attitude to females. On Borneo males stay with their partner for just a few days at most, which is the period of the females' maximum sexual receptivity. Sumatran males may remain with their mates for many months, often until the young are born. This bonding has no sexual implications as mating activity quickly diminishes under these conditions. Sumatran males probably accompany a female simply to defend her as she becomes heavier and less able to protect herself against the resident tigers – a problem, as we have seen, that just doesn't occur in Borneo. Heavily pregnant Sumatran females also have to compete for food against noisy, aggressive siamangs, a species of gibbon not found in Borneo. Siamangs feed in groups and *en masse* they will readily attack all but big male orang-utans if they happen to be competing for feeding rights in trees with ripe fruit. Most orang-utans will back down in the face of a determined guerrilla raid by four or five shouting and leaping siamangs.

This long-term bodyguard relationship may be one reason why Sumatran males seem to be alone in producing courtship displays. For a Sumatran female, choosing a male isn't just a matter of using his genes in the production of new life; she needs to be convinced that he can take care of her during the vulnerable stages of late pregnancy.

Male Sumatran orang-utans use a strange display that involves hanging from high branches, while their long hair untidily cascades down. Based on this show, a female can judge whether this suitor has the strength and stamina to deter a hungry tiger. His appearance is also a factor. A male with well-developed cheek flanges and a powerful long call will add to the overall effect of his display. Males in Borneo have not been seen to display with such extravagance, but this may simply mean that no one has been lucky enough to witness the event. The closed environment of a rain forest makes detailed observation difficult and the cloaking nature of the woodland is one of the most frustrating barriers to our knowledge of the animals that live there. It would be completely wrong to state baldly that males in Borneo never display, just because it has never been seen.

Temporary pair bonding requires a behavioural compromise by both animals. The female's preference for canopy travel is incompatible with the male's liking for low-level or even ground activities. Sometimes they may be feeding in the same tree but the female is 20m (65ft) higher than her mate. The same applies to nesting behaviour. One possible side effect of the long-term courtship behaviour is the possibility that Sumatran orang-utans have become more sociable than those in Borneo. This theory comes entirely from the study of captive animals and we just don't know if it applies to wild orang-utans. A group of Bornean orang-utans in Cologne Zoo had to be split up because of constant aggression, while a Sumatran group of the same size in Zurich Zoo lived in relative harmony. Researchers working at many other zoos report that Sumatran orang-utans of both sexes are far more tolerant of the presence of others than Bornean animals

Young orang-utans may wander a long time before finding a territory they can adopt permanently.

ever would be. There is slender evidence that similar behaviour happens in the wild. A zoologist working in Sumatra once watched fourteen orang-utans feeding from a single fig tree over a five-day period. They showed little inclination to mix but neither did they display the aggression that would have been shown by orang-utans in a similar situation in Borneo.

Female orang-utans give birth around 270 days after conception. Single babies are the rule and twins are extremely rare and quickly die. With an average birth weight of around 2kg (4½lb), labour and birth are relatively swift and trouble-free. All other apes are born into a structured society containing individuals of every age and status from whom they learn social and survival skills, but a newly born orang-utan has just its mother for company and tuition. The normally retiring female orang-utans become almost paranoically shy in the months following the birth of a baby. They shun the presence of all other members of their species and most other animals in an effort to protect the infant. This can be tricky, as other females are intensely interested in young babies and will often try desperately to touch and hold them.

For the first twelve months the baby is absolutely dependent on its mother for everything. As the mother methodically clambers through the canopy, frequently stopping for food, the infant clings tightly to the long hair that hangs loosely on her sides. She sometimes stops to groom and suckle the baby, but only rarely plays with it. This is a major difference from the learning programme of the other apes, whose days are made up of many playtime hours. But as orang-utans are not gregarious animals, is there any point in wasting time teaching youngsters to play or interact with others of the species? The outcomes of meetings with other mothers and babies are extremely variable. Sometimes the two adults studiously ignore each other; occasionally there may be outright aggression or the animals may stay together for a short time. This last scenario has led some zoologists to think that orang-utans recognize their immediate neighbours, even though they may not see them from one year to the next. It is possible that familiar females may be welcomed as company for a while, but strange males would pose a definite threat and must be avoided or driven away, depending on their size. The first signs of weaning appear when the female partially chews bits of her own meal and offers the softened pulp to the baby. Young orang-utans eagerly take solid food but, as is the case in most species of mammal, they are still keen to take milk from their mother for as long as possible.

Juveniles begin to explore and build experimental nests of their own during the second year. With typical adolescent vigour, the youngster spends a lot of its time perfecting climbing techniques and swinging manoeuvres. Around their third birthday, most orang-utans should be building and sleeping in their own night nests, but day and night they always keep very close to their mother. Any time after this point the juvenile is likely to be physically thrown out if it makes any attempt to sneak into a nest and sleep with its mother. By the age of about six, young orang-utans start to wander further afield. Males appear to leave their mother's care at a slightly earlier age than females.

After finally breaking the maternal bond orang-utans become nomads, searching for a suitable territory. They occasionally meet other wandering adolescents and might feed together or play for a short time, but eventually one or all will strike off on the solitary path that dominates the lives of all orang-utans. Once again it is the males that are less likely to stay with others of the species. Adolescents and subadults can lead an itinerant lifestyle for several years before establishing their own home range. Females generally settle down sooner, for they wage few territorial battles among themselves. A young female can be easily absorbed into an existing population. Males, however, have to be very careful to avoid resident adults of their own sex, for these will go out of their way to violently expel intruders. It may be some time before a male becomes big enough to carve out and protect his own forest niche. Succession is quicker if an established male dies suddenly and leaves a power vacuum. There is some evidence that shows this kind of event can bring about an acceleration in the development of local subadults, in a race to seize the chance. There are certainly times when young males display an incredible surge of growth and a rapid appearance of cheek flanges. It could be possible that the presence of a dominant male somehow has the effect of subduing the development of younger males and reduces the competition.

A young orang-utan sucking water from its hair after a rainstorm.

On average the period between births is around five years, but it may be as long as eight years, making the orang-utan reproductive cycle the slowest of all the great apes. A single female may produce only four off-spring in her entire life. Some mammals, such as rodents, give birth to a huge number of babies at one time, working on the basis that if enough are produced there is a good chance that some will survive to breed and pass on their genes. The built-in disadvantage to this mechanism is that the mother cannot possibly give much attention or protection to each of her youngsters; they are largely left to fend for themselves and many of them die early. Bigger animals generally produce fewer young but look after them more carefully and over a longer period. This breeding strategy is a gamble. If the single baby survives to adulthood and produces its own offspring, the effort has been entirely justified. But should it die, everything is lost in one fell swoop and several years may be wasted. Orang-utans, like all of the great apes, will go to extraordinary lengths to protect their offspring because of the investment they have made in their future. Adult females spend most of their adult lives caring for at least one baby at a time; some females may have an active six-year-old while at the same time caring for a new-born baby. With this weight of responsibility it is hardly surprising that female orang-utans are wary and difficult to approach. Their solitary lifestyle still holds many secrets that may be hidden for a long time to come.

4

THE
GORILLA

IN THE AUTOBIOGRAPHICAL *The Voyage of Hanno the Carthaginian*, dating back to 470 BC, there is a passage which probably provides the first known written reference to gorillas. Hanno's task (setting out from what is now modern Tunisia) was to lead a fleet of sixty vessels beyond the Pillars of Hercules (Straits of Gibraltar) with the intention of colonizing West Africa. The book records:

The third day, having set sail, and passed the fiery current, we came to the bay called the Southern Horn. In it was an island, in which was a lake, and in it was another island full of savages, the majority of whom were women, whose bodies were covered with hair, and which our interpreter called gorillas. We pursued these, but could not capture any men; all escaped by climbing up the precipices; but we took three women, who bit and scratched those that overcame them, and whom they would not follow. Having killed them, they were flayed, and we brought their skins to Carthage.

The skins were later displayed in the Melkarth Temple in Carthage under the title 'Gorgadas'. Everyone today is familiar with the appearance of a Gorgon, a fearsome female who has snakes for hair. But this is a modern form of a far more ancient legend. An earlier version is reported by the Greek historian Diodorus Siculus, writing in the first century BC, who tells of a shaggy and enormously strong half-monkey, half-human figure.

After Hanno there is a 2,000-year gap before the apes make a second appearance. In 1625 a book by the English explorer Andrew Battell tells of an encounter with a strange animal in West Africa:

This Pongo is in all proportions like a man; but that he is more like a giant in stature, than a man; for he is very tall, and hath a man's face, hollow-eyed, with long-hair upon his browes. His face and ears are without hair, and his hands also. His bodie is full of haire, but not very thicke, and it is of a dunnish colour.

Young gorillas often sleep together during the midday siestas.

This description is far from perfect but is clear enough to show that Battell had seen a gorilla.

From then on there were endless reports of huge, hairy men blessed with incredible strength and cursed with terrible aggression, but they were dismissed, together with the tales of two-headed lizards and men with heads that faced backwards. Africa was a continent awash with unbelievable stories. Each traveller seemed to exaggerate and dramatize more than the last in order to promote the lucrative lecture tours. The first accurate gorilla description came in a letter, sent along with a skull as proof, from the missionary Dr Savage to the anatomist Professor Richard Owen. Savage was visiting an area that we now call Gabon and there he came across the remains of a beast he described as 'a new species of orang'. This latter term was used then as a blanket name for all known apes; for a time chimpanzees were known as black orang-utans. In a letter of 1846 to Owen, Savage wrote that these animals

are exceedingly ferocious and always offensive in their habits, never running from man as does the Chimpanzee . . . It is said that when the male is first seen he gives a terrific yell that resounds far and wide through the forest . . . he then approaches the enemy in great fury, pouring out his cries in quick succession.

Ten years later the American traveller Paul Belloni du Chaillu recalled what was probably the first ever gorilla hunt using firearms. On seeing the ape he wrote: 'He reminded me of nothing but some hellish dream creature – a being of that hideous order, half-man, half-beast.' These and other outlandish reports of the gorilla's behaviour became the model for lurid press references that helped form the aggressive image that even now has not been totally shaken off. Subsequent travellers and scientists published a stream of tales about this strange animal and all chose to copy the name originally heard by Hanno – the gorilla. In almost every case it was details of the gorillas' incredible size and strength that dominated the descriptions, and most were exaggerated even when concrete evidence was in front of the writer. In 1919 a western lowland gorilla was shot by a French hunter by the name of Villars-Darasse. The animal was reported to have stood 2.84m (9ft 3in) high, but a photograph of the body later published in *L'Illustration*, Paris, on 14 February 1920 showed a gorilla that was judged to be around 1.8m (6ft) tall – and this was because it had unusually long legs. Whatever the hyperbole surrounding the first sightings of the animal, lowland gorillas were not scientifically recognized until well into the twentieth century.

With customary Western arrogance, this is regarded as the 'real' discovery of gorillas, but as always we were thousands of years late. Gorillas had been known and largely understood by local Africans for countless generations. The distribution of gorillas has today shrunk to a fraction of its former area, but just 150 years ago these apes would have been familiar to people living in most of the tropical African forests. As one of the region's largest animals, gorillas were frequently killed for meat. This was obviously carried out on foot by hunters armed only with spears. Gorilla hunting was a formidable task adopted in some societies as a benchmark of a hunter's prowess or even manhood, and there was a tell-tale mark that showed for ever if a warrior's courage deserted him at the last moment. Unless humans do something really stupid or particularly aggressive when coming face to face with a gorilla, the animal will launch into furious and truly daunting displays which are generally all sound and fury. The anger fades away and the gorilla is unlikely to enter battle if the man holds his ground. However, if he turns and flees, the man runs a far greater chance of being attacked. Scars of gorilla bites on the backside or back of a hunter's legs were the ultimate sign of cowardice in many Central African cultures.

Generally, gorillas were treated with respect rather than fear, and some tribal cultures were way ahead of modern Darwinian theory, for the gorillas were seen as strange near-relatives that had merely decided to move away from man and live in the forest; they were looked on as distant cousins. Other peoples viewed the apes as human souls that had returned to earth in a different form after death, and as ancestors they were worthy of deference.

Even after scientific acceptance of this major new primate, the newly enlightened Western civilization, with its liking for sensationalism, did not choose to portray gorillas in a sympathetic or remotely accurate light – particularly as the first reports of their discovery became news at the time when Charles Darwin's controversial theories were causing so much disturbance.

The apes either became entangled in 'evolutionary' jokes or were featured as rampaging brutes in tales from the Dark Continent. All of the early sightings of gorillas were in West Africa, which was rapidly being explored for commercial gain. Much of tropical Africa was largely unknown at that time and there were reports circulating about still more giant apes in the area, but these were largely lost in the morass of exaggerated tales from the Dark Continent. In 1898 a traveller by the name of Grogan was hunting bush elephants in the Virunga Mountains when he discovered the remains of 'a gigantic ape'. He had previously seen chimpanzees and was well aware that this skeleton was far bigger than anything he had experienced before.

The honour of the mountain gorillas' final scientific discovery goes to Captain Oscar von Beringe. In 1902 he was leading an expedition on behalf of the German Empire to meet local chiefs when he stopped off to explore Mount Sabinio in the Virunga Mountains. At an altitude of 2,800m (9,300ft) von Beringe and his companion Dr England encountered some apes. He later wrote:

Of these apes we managed to shoot two . . . It was a large man-like ape, a male, about 1.5 metres [5ft] high and weighing over 200lb [90kg]. The chest without hair, the hands and feet of huge size. I could unfortunately not determine the genus of the ape. He was of a previously unknown size for a chimpanzee, and the presence of gorillas in the Lake [Tanganyika] region has as yet not been determined.

Humanized mountain gorillas of the Virunga Mountains are often very curious about tourists.

The meticulous von Beringe sent back a skeleton which, after some time, was finally labelled the mountain gorilla and given the name *Gorilla gorilla beringei* in recognition of the German traveller. Von Beringe's gorilla was examined by the biologist Matschie, who was fascinated by the classification of animals. Each time a new gorilla specimen arrived from subsequent travellers he would find minor differences between this and existing examples, then announce that a new species had been discovered (his classification system has long been rejected). We can now see that he failed to appreciate the individual physical variations that can exist within a single species. These can be present from birth or can result from life experiences such as accidents and dietary deficiencies. Advancing age brings about its own changes in anatomy that were not appreciated at the turn of the century.

In the fifty years or so following their discovery, hundreds of gorillas were killed by trappers trying desperately to catch animals for the lucrative zoo trade. This was a major new species and would fetch big money on the international market. Live-trapping an adult gorilla is no easy task, but trial and error quickly evolved a failsafe technique. The idea is simple: forget the adults and go for youngsters, as they are less trouble to handle and transport, and adapt more readily to captivity. There is really just one way to take a baby gorilla alive and that is to shoot its mother, along with any other adult that gets in the way. To capture just one infant four or five adults, including the dominant male, would be shot. And all too often the youngster subsequently died of stress or starvation *en route* to its destination, either because it was taken from its mother before it could survive independently or through downright ill-treatment and malnutrition. Newly captured infant gorillas frequently refuse to eat and crude attempts to force-feed them were in many cases just as fatal as starvation.

For decades knowledge of wild gorillas was made up of fragmentary and often misinterpreted firsthand encounters, along with myth and conjecture. No one made a detailed study of their biology and behaviour until an American zoologist, George Schaller, carried out a research project on mountain gorillas in the late 1950s. Observation in their dense forest habitat was difficult in the extreme. In order to see any aspect of their lives Schaller had to spend many weeks tracking timid animals that would disappear as soon as he was detected. Wearing drab clothes, he at first kept his distance to constantly assure the gorillas that he meant them no harm. Gradually they accepted his presence and allowed him closer, enabling him to gain the first insights into the private lives of wild gorillas. Schaller's work has been continued and enlarged upon by many other scientists, including Dian Fossey, an American who devoted more than twenty years to the study of mountain gorillas.

Gorillas are the largest of all living primates and there are three distinct races: western lowland, eastern lowland and mountain. The two lowland races are kept apart by the Zaire river, which, as gorillas do not voluntarily swim, forms an impenetrable barrier. Gorillas deeply dislike water and even when fording a small stream will go out of their way to find a natural bridge, such as an overhanging branch or fallen log, to help them cross without getting wet. They have even been known to turn back rather than get wet.

The lowland races were almost certainly once a single population, but now more than 1,000km (625 miles) separate the subspecies. Geographical isolation and time have brought about minor physical differences. The eastern animals have very black fur compared to their greyer western counterparts; also their teeth and jaws are larger. The mountain gorilla is physically very similar to the eastern lowland, but with even longer jaws and teeth. They also have broader chests, together with wider feet and hands. The forearms of the mountain race are shorter than those of the other two and their body-hair is longer; the latter is probably an adaptation to cope with the cooler temperatures found high in the Virunga Mountains. The word 'virunga' comes from a local expression meaning 'isolated mountains that reach the clouds'. Despite being so close to the Equator, these cloud forests can become very cool at times, resembling a temperate rather than Equatorial woodland.

The physical and behavioural differences between the three gorillas are so minor, it is likely that they split off in relatively recent times. All sub-

This male western lowland gorilla has shorter hair than the mountain gorilla and a redder coat than the eastern lowland animals.

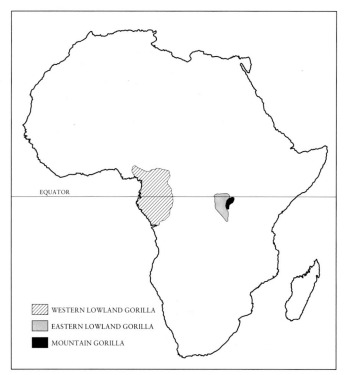

Fig 4
The distribution of gorillas

WESTERN LOWLAND GORILLA

EASTERN LOWLAND GORILLA

MOUNTAIN GORILLA

species are essentially terrestrial and quadrupedal; of the three, mountain gorillas spend the greatest percentage of their time on the ground. Females and juveniles regularly explore trees, where they climb slowly but surely. Youngsters move from branch to branch by grasping with their long arms and swinging in a way that is similar to the brachiation of gibbons, although it is a lot less rapid and agile. Fully grown males will venture into trees only if they contain food that is worth the effort to reach.

Even in the tropics, trees have fruiting seasons. Some bear a harvest only for a short time and there is competition for the privilege of eating it in prime condition. Some fruit is popular with birds, apes and other mammals, so a mature, ripe crop soon disappears. Heavy males will carefully and slowly scale trees up to 30m (98ft) high in order to reach rare and favoured fruit. Because of their dominant status, they have first option on these choice morsels, but they are severely hampered by their weight. Juveniles and females can venture further out on thin branches to pick fruit that is unreachable by heavy males. Skeletal analysis shows that around 20 per cent of gorillas have suffered, and recovered from, broken bones that could have been caused only by falling from a tree. Gorillas are wary climbers and use all four limbs when above the ground. They rarely swing from branch to branch suspended only by their forearms – a fast way of travelling often used by chimps and gibbons. Neither are they happy about jumping over anything other than very short distances. Gorillas obviously feel most secure if all four limbs are in contact with a hefty, solid branch.

When coming down from a tree, gorillas descend in reverse, hindfeet first, carefully checking hand-holds and branches all the way to the ground. As they get older and heavier, they increasingly spend more time on the ground, where they walk on all fours. Their back feet are placed flat on the ground like a human foot, but the weight on their forelimbs is taken on their knuckles, with the thumb outstretched backwards or sideways. The first two joints of each finger have a thick-skinned callus on which the animal's considerable weight is rested.

Gorillas live in extended family groups which range in size from two to thirty-six, with an average membership somewhere between five and ten. The make-up of the group will fluctuate with births, deaths and desertions. One group found comprised just five males, which goes to prove that the permutations are almost boundless. Whatever the make-up, groups are always controlled by a large adult male known as the silverback, named for the grey/silver band of hair that runs down his back, sides and flanks. Apart from the obvious natural differences in size and colour among individuals of different ages, gorillas can be identified through the shape of their nose. The folds, wrinkles and outline of a gorilla's nose are just as distinctive as a fingerprint (which they also possess) and researchers soon learn to differentiate animals through noseprints.

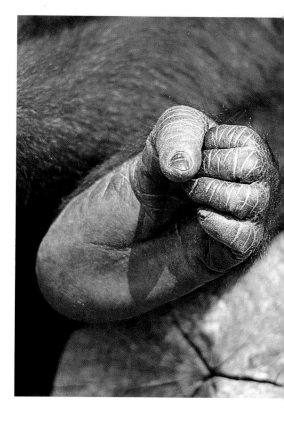

The hindfoot of a lowland gorilla.

Gorillas can be identified by their distinctive noseprints.

There may well be more than one silverback in a group, but when this happens one will always occupy the unchallenged role of leader. Schaller discovered one group that boasted four resident silverbacks and they all appeared to live together peacefully. Should the dominant male become injured or lose his strength and faculties through age, he may be challenged and ousted by a younger male. There is only a slight chance of the transition being really violent; occasionally the ex-leader is forced to leave, but a transfer of power may equally well take place with few ripples. A deposed male may remain in a group that he once led. Relationships between two silverbacks in one group are remarkably peaceful. One researcher showed in a well-documented study group that a lower-ranking male was inclined to avoid the leader by sitting or eating furthest away from him. But even this mild behaviour has not been noticed in other groups.

A saddle of silver hair visually marks the dominant male as the nucleus of the group and it is his presence that bonds the apes into a workable social unit. Should the lead male die leaving no immediate successor, the family ties break down and individuals wander off to join neighbouring groups. However, family loyalty is the backbone of gorilla social life and it takes disease or violence to remove a male from his family. As gorillas are known to live in excess of forty years, in the wild a group can be controlled by one individual for more than two decades. Silverbacks are highly protective of their groups and will occasionally die in their defence. Without the male, the remaining gorillas frequently refuse even to defend themselves against attack and poachers readily use this knowledge to aid their grisly trade. Silverbacks are the first to be killed; with him dispatched, the poachers are safe from attack and the other gorillas become so distressed and confused that they are easily captured or killed.

Silverbacks lead the family group in most of the major decisions. They decide where the group will feed, where they will sleep and in which direction they will wander. As undisputed boss, the adult male enjoys a peaceful life, largely undisturbed by friction within the group itself. Squabbles are confined to juveniles and females; the male appears to be uninterested in them unless the quarrel gets out of hand. He then moves in to restore peace with a swift cuff or warning glance.

A typical group would consist of the silverback, four adult females and four youngsters of assorted age and gender, but this arrangement is almost infinitely variable. There is a loose hierarchy in each group, with the large silverback ranking highest. Beneath him, in order of rank, come the other silverbacks (if present) and the blackbacks – males and females are approximately on a par, although this does depend on their own personality to a degree. Older females usually hold seniority over younger peers and females with young always outrank those without. Next come juveniles and finally infants. In practice the hierarchical system means that lower-ranking animals step aside when a high-status individual wishes to pass, and also high-ranking gorillas tend to have first choice of favoured foods.

To minimize inbreeding, on reaching puberty most females leave their birth group and join another. This avoids the genetic problems created by closely related animals mating. Females do occasionally stay put once they reach sexual maturity and could possibly mate with their own father. The choice of staying permanently is made by the female herself, rather than by an overpossessive silverback who refuses to relinquish her. The decision to stay is probably based upon the quality of the silverback. The female must decide if he has the skill to lead the group to good feeding sites and is strong enough to defend her and any future offspring against attack. If she feels that a silverback cannot fulfil these roles, she moves on to another group and another male.

It is these female transfers that bring about the most aggressive confrontations between dominant silverbacks of neighbouring groups. Actual fighting is rare but both males become very vocal while the female moves from one group to the other. Tree shaking and chest beating occur in both camps, and all members of both groups become very agitated, joining in with hoots and screams. Once a final decision has been taken, the female generally stays with that group for life. She will probably then be one of a number of females who have arrived by the same route but from different groups. It is possible that one or two adult females may be related, but most will have no blood-ties and this produces a healthy breeding group.

Males may also opt to leave the family group, but for completely different reasons. As the male to female birth ratio is about 1:1 and breeding is conducted on a harem system, there are many males who have no access to mates. Adolescent males stay with their birth group until they reach full size. Although they have not yet acquired the distinctive silver saddle, fully grown blackback males are biologically programmed to find females. But the intimidating presence of the dominant silverback prevents free mating. Unlike most other social animals, gorillas rarely enter status fights within their own groups. A young pretender accepts the superior position of the silverback without challenge until one day he just chooses to leave. Unlike females, males do not need another group to join as they are perfectly capable of defending themselves against all likely enemies. Males will wander off alone and can endure solitary lives for several years; they sometimes meet and join forces with other itinerant males and form a loose alliance. But, more likely, a lone male will unite with a transferring post-pubescent female and together start an entirely new group. For some time after reaching full body-weight, males retain their juvenile dark coloration and are known as blackbacks; the silver patches begin to appear when the gorilla reaches about fifteen years old and it extends with age. Excited reports have been made by people who claim to have seen pure white gorillas; in reality they have probably seen a grizzled old male whose silver coat had spread to cover his entire back. This is particularly common in the western subspecies. White gorillas are not completely unknown: an albino by the name of Snowflake lived in Madrid Zoo for many years and was one of the most popular attractions.

In the course of constant wandering, gorillas regularly cross the paths of other groups. They are peaceful animals and trouble is averted through the noisy rituals that keep the groups apart. They seldom wish to risk the potentially violent outcome of confrontation. Most of the time they show no desire to make contact with opposing groups and direct meetings tend to happen accidentally. After initial hostility when two unfamiliar groups first meet, tension lessens with time. When the two subsequently meet, a wary truce is likely. There is some evidence that neighbouring groups not only recognize each other but occasionally actually welcome contact. Schaller tells of the time he witnessed a meeting between two groups: they must have been well acquainted, for gorillas from both camps slept together in the same nest.

Wandering unaccompanied silverbacks are one of the few causes of real violence between gorillas themselves. When an entirely unknown single male meets a family, the group silverback treats the encounter with great suspicion. Tree shaking may be the first response to the intruder; this is a common status display among primates. Rather than launching into direct physical combat and risking mutual injury, the apes initially show their power in more subtle ways that avoid actual contact. Adult males can shake larger trees than other gorillas, and this shows they are stronger. When there are no big trees or branches available, silverbacks will tear up shrubs and throw them as a way of demonstrating their strength.

Tree shaking is a ritual closely watched by all gorillas, even if they are not involved in the dispute. In this activity they can read the subtle signs that advertise the current prowess of the silverback. And the messages are all taken in by the new rival. If the tree performance is less than impressive and fails to drive away the enemy, the silverback goes into a more formal display. This often starts with a series of short hoots becoming louder and quicker until they reach a screaming crescendo. For some unknown reason the male often pauses for a moment to hold a leaf between his lips before continuing. After calling, he will then rear up on to his short back legs and, with cupped hands, rapidly beat his lower chest. At his fastest, a gorilla beats at the rate of ten slaps per second, producing a sound – and speed – that could be matched only by a gifted drummer.

Gorillas also hit their thighs, shoulders, other parts of their body and even tree trunks when really annoyed. Humans always use their fists when attempting to mimic chest beating, but no gorilla would do this, as cupped hands make far more noise than fists. Beating is programmed into the gorillas' genetic make-up; it is performed by captive, human-reared youngsters who have never seen an adult display and therefore could not simply be copying the action. All members of a group will beat their chests when excited but none with the powerful resonance of the large males. It can sometimes be used in non-aggressive contexts, possibly as a form of communication or, in juveniles, as play.

Chest beating is frequently accompanied by coughing grunts and loud roars. Gorillas appear to have the ability to recognize the individual owners of voices, just as we can. Prolonged calling carries up to a mile in the forest and is probably a way of establishing personal credentials and avoiding trouble before matters get out of hand. At times of extreme excitement, males make short sideways dashes, picking up leaves, twigs or whatever comes to hand and scattering them around. At this point other gorillas rapidly move away, as an angry silverback will hit out at anything in his path when charging.

When a display reaches fever pitch the gorilla usually draws back his lips to expose his teeth and the culmination of the show is a powerful forehanded thump on the ground. In the face of a real enemy, other group members join in the display, adding their own screams and charges to the overall mêlée. Even the juveniles may copy the chest beating, branch tearing or ground thumping, but only the silverback completes the full ritualized exhibition. A screaming group of charging gorillas produces a level of sound that has to be experienced to be appreciated. The noise is appalling. With this team on his side, the incumbent silverback has a psychological advantage and it all helps to intimidate a lone stranger. This has the effect of ensuring that only the strongest and most confident males break through the defence and get a chance of challenging for leadership.

A full display, with all the subtle status nuances that are hidden from human eyes, is enough to solve most silverback disputes. But when two males are evenly matched and their displays are equally powerful, fighting may break out and this can be bloody. In real combat gorillas use their long canines as the prime weapon. Biting like dogs, they are capable of inflicting terrible wounds which may result in permanent injury or even death. Dian Fossey discovered that up to 75 per cent of silverback remains showed signs of head wounds and an even higher percentage had missing canine teeth. Both are probably caused by inter-male fighting. She also reports having found two skulls that contained pieces of canine teeth still embedded in the bone.

Should the resident silverback be killed, the victorious stranger attempts to take over – and not always successfully. The sudden change of leader may be willingly accepted by the remaining members of the group or they may choose to reject the interloper and move on. Those that stay are putting their existing offspring at risk, for newly installed males often kill young infants. This behaviour is not confined to primates; other colony dwellers such as lions are known to do exactly the same. Each male unconsciously wishes to pass on his own genes to the next generation. Genetically this makes perfect sense, for he has just deposed the ex-ruler and therefore may be a stronger animal who is capable of producing bigger offspring. If the new male is forced to wait until the female is ready to breed naturally, his potential for siring offspring is reduced. By killing the existing young, the male encourages his females to become receptive sooner and accelerates the speed with which his genes will infiltrate the group. Infanticide by silverbacks accounts for a large proportion of deaths in gorillas below two years old.

In the harem system, where one male dominates several females, size and strength play a significant part in the breeding success of males. The largest males mate more often, therefore, passing on their genes to produce another generation of large males, and so the cycle continues. This inter-male competition is one possible reason for the marked sexual dimorphism (structural difference between the male and female) in gorillas. The male, at 150kg (330lb) on average, is twice as heavy as his females. Another explanation may be that males have evolved their present size in order to protect their family group against predators. As terrestrial animals gorillas are vulnerable to attack from powerful carnivores such as leopards. One of the duties undertaken by males in many colonial species is group defence, and they tend to be larger and faster as a result.

Like all of the three subspecies, mountain gorillas live in groups made up of individuals of all ages.

Life within a well-integrated social group has an evolutionary advantage for each individual member and the species as a whole. Youngsters learn many necessary life skills by copying not only their parents but all other apes in the group. Gorillas have a slow reproductive rate and youngsters are totally dependent on their mother for a long time. Alone, nursing females would be vulnerable, but within the safety of a group they reduce the chances of being singled out by a predator, and have the added benefit of a formidable bodyguard as protection.

Not only are males heavier, they are also armed with longer canine teeth. To power these teeth, males are equipped with strong temporal muscles that wrap around the sides of the skull and are anchored on to a tall, keel-shaped bone at the top of the head. This is known as the nuchal crest and it is what gives the males their characteristic oval-shaped head that is so very different from the females' more rounded head. At times of excitement the male raises a narrow strip of hair on top of the crest. This is known as pilo-erection and makes the head look even bigger, more oval and therefore more masculine and threatening.

Day-to-day travelling in gorilla groups tends to be unpredictable and slow most of the time, but the animals can move swiftly if food is scarce or when they are exploring unknown country. Cast-iron rules of travelling are impossible to formulate, for here, along with many other types of behaviour, I can offer only generalizations. Unless they are disturbed, the gorillas' day starts shortly after dawn (around 6.00 a.m. in the tropics) and, following the enforced fasting of night, the morning consists of extensive feeding. This is when the majority of their food intake takes place. They then rest for a few hours in the middle of the day before feeding again in the afternoon and early evening. The time when gorillas are most likely to travel is immediately after they have woken from the midday sleep.

A lack of enemies, abundance of food and their large size all combine to make travelling unnecessary for gorillas. Their daily range is relatively small, but they may wander over a wide area in the course of a year. The distance covered in any period depends on several different factors, such as the weather, competition and food availability, but it is a rare day when a group covers more than a couple of kilometres. They are perfectly capable of travelling much further in that time, but unless there is a real need for speed, gorillas prefer to take life at a slow pace. A feeding group is also very noisy. As they move gradually through the forest, they have no reason to maintain silence and even a small group of six can sound like a much larger number.

Heavy rain is guaranteed to interrupt most activities. Light showers are ignored but the torrential downpours that lash the tropics make even the gorillas pause. Once the rain becomes too heavy for comfort, they stop in their tracks. They make no real attempt to find cover, even though it may be nearby. They just sit the rain out and, from a strictly human viewpoint, look wholly uncomfortable about the rotten weather. Their hair may be thick but it is not dense enough to give real protection against the rain. The apes soon become thoroughly wet. But tropical rain appears and disappears very quickly, and once it has passed the gorillas return to feeding and, after a while, they often begin to steam slightly.

Around midday, when the sun – if visible – is at its highest, the group stops for a rest. The male chooses his spot and the others lie down in a loose group around him. If the undergrowth is long, each gorilla tries to beat it flat around the chosen sleeping area. This makes a more comfortable bed and also gives each animal a good view of its surroundings. While watching gorillas sleep at noon, it is tempting to draw parallels between their behaviour and the human habit of taking siestas in warm climates. The gorillas sit and doze at first. Then a spot of grooming might take place. Unlike most other primates, each gorilla takes care of its own toilet routine; the mutual grooming that is so common and important in the social life of chimpanzees is quite rare among gorillas. The one exception occurs when a female cleans an infant who is too young to carry out this chore for itself. The youngster is held in one hand and groomed with the other. Attention of this intensity is welcomed by infants, particularly when they have been largely ignored during the feeding over the previous hours. Females occasionally groom silverbacks during rest periods, but true grooming between adults of the same gender is unusual.

Cross-grooming in other primates is used not only to keep each other clean and tidy but also as a way of strengthening bonds and minimizing tension among adults. As gorillas have a more stable social life and are non-aggressive to members of their own group, the need for regular mutual reassurance is less important than in more volatile species, where fighting may break out unless the group is kept constantly pacified. Gorillas appear to groom merely for the sake of hygiene. For a while the apes carefully search and comb their hair, pulling out parasites and debris that have accumulated during the day. The adults then slowly drift off to sleep, while the juveniles play.

There are no set rules for games among gorillas. They spend their time climbing, swinging on branches and vines, wrestling each other and playing with sticks and anything else that can be found. I watched one rolling over and over down a slope. On reaching the bottom, he scampered back up the bank and started somersaulting all over again. Young males show a tendency to practise chest beating in play from a very early age; they have been seen drumming experimentally, and weakly, from the age of just four months. Obviously these early efforts bear little resemblance to the polished, near-deafening display of an adult, but practice makes perfect and so they continue. Very young gorillas never venture far from their mother and return to her frequently, almost as if they need to reassure themselves that she is still nearby. Older juveniles include the slumbering adults as toys in their games. Females and males, including the silverback, are all unwitting participants. They are poked, pulled and prodded; climbing on adults is a popular game that features in all midday rest-times. Gorillas show more tolerance than many human parents would under these conditions. Only rarely do they exhibit irritation at the constant disturbance and this is communicated either through a direct look or a push with the hand. This subdues the youngsters for a short time, but they soon forget the warning and get back to the job of tormenting their parents.

Eventually even the boundless energy of the juveniles is sapped and they too settle down. Well-fed, undisturbed gorillas at rest show their feeling of well-being with low, throat-clearing sounds that Dian Fossey describes as 'belch vocalizations'. One animal might give out a contented cough and, if conditions are right, others soon join in the quiet, slow chorus and the overall communal effort probably makes the apes feel even more relaxed. It is this sound that is used by trackers and researchers when they are trying to reassure the gorillas that they mean no harm. The evocation of pleasure has a pacifying effect on the apes and can cool down potentially sticky situations. From a strictly human viewpoint, it really does sound as if the gorillas are happy with the world. In early afternoon the gorillas, signalled by the activity of the silverback, stir, stretch and move on. During feeding they may lose sight of each other, but there is no danger of getting lost due to the coughing 'barks' that come from each animal to indicate its presence to the others. They eat and wander until shortly before dusk, when the group again gathers close together ready for sleep. Gorillas seem to have no obvious preference for the site of a night camp; they just stop near to their last feeding ground.

Like the other apes, gorillas spend their nights in individual nests. These can be built either directly on the ground or in the lower reaches of a tree. The position is dictated by local conditions, individual preference and the age of the gorilla. Often in one group there is a mixture of both ground and tree nests. The exact choice of site is controlled to a large extent by the surroundings; some habitats offer ready-made nests. In bamboo forests there are more nests built off the ground. Thick clumps, measuring more than 2m (6½ft) across, can quickly be modified to provide a solid sleeping platform. A few snapped stems and added branches and the gorilla has a strong bed for the night. Sometimes they avoid sleeping at ground level for practical reasons. Lowland gorillas often inhabit forests with wet floors; this is an incentive for building in trees.

At its most basic, a nest is built on the ground and comprises a wide circle of vegetation pulled in towards the gorilla, who sits in the centre. This forms a thick rim of compressed plant material around the edge which thins towards the middle, giving an almost doughnut-like appearance. Where the animal actually sits there is often no added bedding, only the vegetation that happens to be growing there. Interestingly, gorillas hardly ever incorporate food plants into their nests, even though they will add extra bedding from non-edible plants. Thick branches are intentionally twisted or sharply bent to break the stems and prevent them springing back into shape. A nest is not a complicated structure, so there is no real building skill involved. The gorillas do not attempt to weave or layer the material in any way; the nest is simple and may take only two or three minutes to build.

Nests in trees are usually built in forks, which help to support the sleeping animal. Torn-off branches are added for extra strength and to act as padding. Tree nests have been found up to 8m (26ft) above the ground. For obvious reasons, it is the lighter juveniles that usually build in trees, while the more solid males prefer ground sites, but even this is variable. Lowland gorillas build more raised nests than their mountain cousins, and individuals in all three races, just like humans, have their own foibles when it comes to picking a bed. The precise function of ground nests has long been the subject of discussion. They offer no protection against the elements and add little in the way of extra comfort to the sleeping ape.

Ground-nest building is possibly vestigial behaviour left over from a time when the gorillas' ancestors spent most of their lives in trees. Orang-utans are a modern example. Their sleeping habits show that a vital safeguard needed by an arboreal animal is a secure platform which will support the weight of a heavy body throughout the night. Gorilla ground nests are probably now built from genetic habit rather than of necessity. Sometimes gorillas sleep directly on the ground and make no attempt to construct a bed. This could be a sign that slowly the species is abandoning the custom now that their other behaviour has altered.

It is in late afternoon that the male picks the night nesting ground. Once he starts pulling vegetation into a bed the others gradually stop feeding and settle down. Gorillas usually sleep in a foetal position, resting on their sides or fronts with arms and legs tucked close into their body. They have also been seen sleeping upright, resting their backs against a tree, like a human dozing in an easy chair. The family spreads out in a rough group, with the silverback usually, but not always, near the centre. Individual nests radiate out from the male in order of status, with high-ranking gorillas closest to the silverback and juveniles confined to the edge of the group. High-ranking gorillas leave wide gaps between their nest and that of their neighbours, while lower-ranking animals prefer theirs much closer together. These formulae make the gorillas sound slaves to habit and this is misleading. Endless variations can be introduced to this 'basic' arrangement and ten consecutive nights can produce ten different nesting patterns. Gorillas are sophisticated animals and show a range of individual preferences that vary from the accepted norm. Their actions can be brought about by something as whimsical as mood. Gorillas experience anger and frustration and will even sulk if rebuked or displaced by a senior. Disputes that flare up during the day may be carried on into the evening, when the antagonists deliberately settle on opposite sides of the camp. Adult males have been known to sleep at the very edge of the group, because they appear to like the position. It is dangerous to say that all gorillas have one kind of behavioural pattern just because most of them display it.

By the age of about two and a half, gorillas are starting to build their own nests. Before this they sleep with their mother. Night nests remain visible for many months after their construction, due to the flattening effect the gorillas have on the nearby vegetation. This is extremely useful for researchers trying to estimate gorilla populations, when the animals themselves may be too wary to be approached directly. Even the number of dependent young can be accurately calculated by looking at the droppings. As a result of their diet, gorilla dung is solid and fibrous, and it takes some time to degrade. Gorillas frequently defecate in their nests and over a period of days it is possible to 'read' these remains and build up a detailed picture of the group composition. The droppings of an adult female are approximately twice the size of those of an infant, so nests containing both obviously accommodated a female and her youngster.

The deciphering of droppings is complicated by the fact that gorillas sometimes practise 'coprophagy', in other words they eat their own dung. Captive animals have always been known to do this and it was assumed that the behaviour was exaggerated by the false conditions and boredom. In fact, animals eat their droppings for several reasons, one being a mineral deficiency in their diet; some elements are not readily absorbed and may require a second passage through the digestive system before they are assimilated. This may be a possible explanation, but both wild gorillas and those kept in decent zoos receive a well-balanced and healthy diet, so serious deficiencies should be rare in both cases. Boredom could be a valid explanation even in wild populations. Dian Fossey reported that coprophagy is at its most common during the rainy season, when gorillas are forced to sit still for long periods. Eating droppings could be a way of passing the time, rather than revealing a dietary problem. One possible exception to this theory is the female's habit of eating her infant's droppings during the first six months or so of its life. Nursing mothers pass on valuable vitamins and minerals to their young through milk. Not all are absorbed by the infant's body, so it may be of use to the mother to eat the droppings in order to take in whatever is left.

Babies are the focus of a huge amount of interest from the rest of the group. It is not only the females that wish to touch and handle infants; gorillas of both sexes and every age are very curious about new arrivals. The reason is that gorillas reproduce at a slow rate and births are far from regular events. Under ideal conditions the inter-birth period, which is the entire cycle from birth, through suckling and total infant dependency to the next birth, takes around four years for each female. This time is obviously shortened if a baby dies not long after birth.

Gorillas suffer an average mortality of about 40 per cent in the first three years of life. This figure means that each female produces one surviving youngster only every seven years – a painfully slow breeding cycle. The formula is complicated by the fact that young adult females show a tendency to transfer to other groups, while older females are more settled. Some will move while still caring for very young babies, which inevitably increases the risk of the infants being killed by the resident silverback. Age brings stability. Mature females may have a colourful past but eventually they settle down and stay with one group and their breeding pattern becomes a little more predictable.

Physical watersheds are well documented in zoo animals as their ages are exactly known and reams of research material have been published on the life history of captive gorillas. The arrival of puberty is just as erratic in gorillas as it is in humans. Sexual maturity in females is not difficult to pinpoint, for this is marked by the beginning of menstruation, which first appears on average when the animal is around seven years old. Ovulation is visually signalled by a noticeable swelling of the skin around the vagina. Females have a regular cycle of about four weeks but this regularity may not appear for a year or so after puberty. During this uncertain phase, females may mate but not necessarily conceive; they might not produce young for two or three years after sexual maturity. Full puberty in males is less easy to pinpoint, but they have been known to father young at seven years old in captivity. One indication is the sudden acceleration of growth. Until puberty there is little difference in the size of males and females, but when the testosterone kicks in, males put on a lot of weight. As this takes some time to show, the growth spurt tells us that puberty has been and gone, not that it is in progress.

Much of our detailed knowledge of breeding behaviour comes from observation of captive gorillas; early researchers thought it unlikely that this was very different from the timetable and habits of animals in the wild. Information gleaned from studying zoo animals can be very misleading, and breeding is a good example of how wrong conclusions can be drawn if we assume that behaviour is identical in the wild. Gorillas are comparatively rare in zoos because they are expensive to buy, house and feed. Few collections can afford to have more than a tiny number. Often there is just one pair that have been together since they were young. Under these circumstances the male has no competition for mating rights and no powerful silverback to drive him off when the female is receptive. Almost certainly zoo males father young at an earlier age than they would within a large group in an African forest. It is physically possible for a male to breed successfully at seven years old, but in the wild they would have to wait until the age of at least ten or even twelve before reaching a size and seniority that gave them the opportunity.

Within an average thirty-day cycle females are at their most sexually receptive for about three days in the middle. During this oestrus period there is a subtle change in the behaviour of both the female and nearby adult males. Shortly before oestrus, the silverback can be seen sniffing the female's rear end for signs of receptivity. The female responds by staying close and staring directly at him. Copulation is usually invited by the female, who backs up to the male with her rear in the air and her shoulders on the ground. She will often then peer round to look at the male. Copulation is brief and takes place many times during the oestrus period. A receptive female will most often mate with the dominant silverback if he is nearby, but failing that she will approach any group male. Whether or not the female presents directly to them, all group males show interest in receptive females. If there are signs of mating with a subordinate male, the silverback is likely to stop this by physically separating the two in a less than gentle manner. This deters other males from taking liberties, at least in the presence of the silverback.

The length of pregnancy in gorillas is yet another figure difficult to measure exactly. They mate so often during oestrus and on into pregnancy that it is impossible to know quite when conception takes place. One estimate, based on zoo animals, shows a gestation of anywhere between 237 and 288 days, with an average of 258.4 days. As humans show a wide range of durations in perfectly healthy pregnancies, it should be no surprise that gorillas do the same. Pregnancy does not last an exact number of days but falls within a fairly elastic period.

Because their habitat does not go through drastic annual changes in weather or food supply, gorillas do not have a set breeding season; they can conceive and give birth at any time of year. The birth itself is relatively quick compared with a human delivery, because the size of the baby's head is much smaller in comparison to the mother's pelvic diameter. From start to finish the process takes less than an hour unless there are complications. The majority of births take place in night nests. Dian Fossey's work shows that females eat the placenta following the birth of a live infant but do not if the baby is born dead. The eating of such a rich food source is common among many mammals, but it is interesting to speculate why placentas are rejected after a stillbirth. It may be that, using some innate knowledge rather than intellectual reasoning, the gorilla is aware that a stillbirth is caused by some physical problem and the placenta may not be a healthy meal. At birth baby gorillas have pink skin, but this odd-looking hue disappears after a few days of exposure to the sun and then becomes matt black like all other gorillas.

Single births are the rule for gorillas. On the rare occasions when twins occur they are usually too small to survive for long and one or both soon die. New-born infants weigh around 2kg (4½lb), one-eightieth the size of a silverback; in humans the ratio of a new-born baby to an average adult male would be around 1:26. Following birth the mother carefully licks all fluids from the baby's sparsely furred body. What happens next will then depend upon the gorilla's age and experience. Females who have had two or three youngsters hold the babies gently, supporting their bodies while the infant takes milk. First-time mothers often carry their babies around in less suitable ways. They have been seen dragging them by one leg or holding their head while the body swings freely. Some show no sign of encouraging the baby to suckle; in fact, they appear to have absolutely no idea of what to do with the strange little creature. Infant mortality is slightly higher in first-time mothers than for more experienced females, so they do seem to learn as they go on. When a baby dies some mothers will carry around the body for several days before losing interest in the silent, non-responsive object. Some cling on with grim determination until the corpse falls apart through decomposition.

Research in zoos shows that females who have been brought up isolated from other gorillas have the most problems adjusting to mother-hood. One explanation for this ineptitude is that first-time mothers have not seen how other, more experienced females handle, feed and groom their babies. Another possibility could be the result of the gorillas' willingness to partially share responsibility for their young within the group. When the group is relaxed, an adult female will often gently take away an infant from its mother during the midday rest break, giving it cuddles and prolonged grooming in her own nest. This is known as 'aunt behaviour' and happens only if the mother chooses to hand over her young and if the baby itself is happy to go. Not all females are allowed this privilege; some are pushed away without getting the chance to hold the youngster. When feeding, travelling and particularly in moments of stress, the mother swiftly takes control of her own baby.

There are several theories about this baby-sitting behaviour. First, the real mother is temporarily relieved of her duties, giving her the chance to relax for a while. This welcome release from pestering will be familiar to anyone with young children. Secondly, it introduces the infant to the fellow group members who will play such an important part in the first decade of its life. And, finally, inexperienced adult females are given the opportunity to handle young, before they take on the job for real.

At birth young gorillas show about the same physical development as human new-born babies. They are weak, lack motor-control and their senses are unrefined. They do, however, have one built-in skill that is essential to survival: a powerful grip that helps them cling strongly to their mother with both hands and feet. The ability to support their own weight while hanging from an overhead rope, using a single-handed grip, is well known in human babies, but it disappears shortly after birth. The initial powerful reflex quickly disappears and takes many months to return, while in gorillas the strength steadily grows from the first day. A resting female carefully cradles her young most of the time, but she cannot fully support it while feeding or walking. Even the youngest dependent babies have the strength to hold on to their mother while she slowly but steadily moves through dense undergrowth. They are obviously less of a problem for the adult and are also much more likely to survive if they take care of their own clinging mechanism. Very small youngsters are carried or hold on to their mother's chest – a position that protects them from external dangers and gives easy access to milk. As they grow and become heavier, this dangling position gets more awkward for both animals, so gradually the infant spends more time riding on the adult's back.

Gorillas might be comparable to man at birth but their immediate development is far more rapid than ours. Co-ordinated reaching and grasping actions appear at around nine weeks, shortly afterwards solid food is eaten for the first time and then the infant sits unaided. From around fifteen weeks old the young gorilla begins to walk on all four legs like a miniature adult. Other juveniles are fascinated with this potential new playmate, but their overtures often send the infant darting back to the safety of its mother's arms. Confidence quickly grows and after a few days the baby is wrestling along with the others. Play is not just a pleasant way of passing the time. Mock fighting and chasing sharpen the reflexes of the youngsters and build up their muscles. Play-battles help each of them evaluate the strength and aggression of their peers, a knowledge that will go to form the basis of the hierarchical 'pecking order' that is so important in later life. Gorilla infants also spend a lot of time on their own, exploring and learning about their surroundings. They investigate sticks and leaves, perfect their climbing skills and, by closely watching others, absorb a great deal of information.

The first few years of life are a very dangerous time for young gorillas. Being so small they are vulnerable to a wide range of predators. They do not have the natural escape speed of many young prey animals: a two-year-old gorilla, for example, can be outrun by a healthy adult human. But they do have the considerable advantage of the group, and its silverback, as protection. As the sexes are born in roughly equal ratios, it appears strange that adult groups consist of far more females than males. The reason is twofold. First, at any one time there will be a fair number of solitary males that are not part of any group; and, secondly, the mortality of males is higher than that of females throughout their respective lives, as it is males that run the risks of death and injury in their role of group defender.

The dense nature of the gorillas' habitat makes accurate assessment of their population a nightmare. Mountain gorillas have been thoroughly and intensively studied over the past three decades and in the early 1990s it was believed that most major groups had been found, producing a world total of around 360 animals. Then another 300 were discovered living in the appropriately named Impenetrable Forest in Uganda. Lowland gorillas are scattered across a far wider area, in even more inaccessible habitats. Current estimates of their population are just that – estimates. Eastern lowland gorillas are thought to number around 10,000; the western lowland population is believed to be somewhere around 100,000. But it has to be emphasized that these are not substantiated figures, the exact numbers could be very different – in either direction.

Baby gorillas are tiny compared to the adults.

5
MAN:
The Super Ape

Nder the linnaean system of animal classification, each species is grouped together with relatives within a family. Animals that are closely related are listed under the same genus; they are then given a second name which identifies the individual species. The words used are frequently strange hybrid mixtures of Latin and Greek, with the names of their discoverers sometimes thrown in for good measure. The idea is to provide a universally recognizable name which avoids confusion when scientists of different native tongues get together.

The names are chosen almost whimsically by zoologists rather than linguists and they usually refer to some aspect of the animal, either physical or behavioural, that makes it different from other creatures. For example, the polar bear is listed as *Ursus maritimus*, which translates as sea bear. When we came to classify humans, with typical immodesty, we honoured ourselves with the name *Homo sapiens* – wise man. As we were the only animal on earth to call ourselves anything at all, it was concluded that we were the most intelligent. Linnaeus himself wrote: 'I am well aware of the vast difference between man and ape, when considered from a moral viewpoint. Man is the only creature whom God has blessed with an immortal soul.'

With a more detached and less emotive view, Desmond Morris reviewed the original choice of name and suggested that humans should be redesignated as the 'naked ape', which is a name zoologists from another planet may have chosen for us, if they had been given the task. When it comes to biological classification, the relationship between the great apes and humans lies in the differences that separate us, but from a personal point of view I find it is far more interesting and illuminating to look at the things we all have in common.

In pure evolutionary terms, humans are one of the side branches on the ape family tree. There are minor anatomical features that differentiate us from the others. Our arms are shorter and our legs longer in relation to our body size. We walk upright and lack an opposable digit on our feet. Our canine teeth are smaller and our bodies are considerably less hairy and muscular. There are other distinctions but these are mostly hidden. The similarities outweigh the differences by a huge margin.

Many biologists believe that humans are separated from the apes by

Irrigated belt around the River Nile in Egypt. Simple water-management techniques enabled humans to grow food and thrive in areas such as deserts, where no other primates could survive.

what is found in our heads rather than by other parts of our anatomy. For both scientists and philosophers it is the brain that makes us distinctly human. Physically we are slower and weaker than the anthropoid apes; we are far less able to defend ourselves against predators and, without the aid of tools in some form, only a tiny fraction of our species are now equipped with the skills to comfortably survive in any natural habitat. The skulls of early man had short nasal clefts, showing that even then our species had a poor sense of smell. Human teeth are small and puny, unsuitable for either shearing meat or constantly chewing up leaves and grass. Humans are physically one of the world's most spectacularly non-specialized animals: we can't run quickly, climb trees confidently, crack nuts with our jaws, migrate efficiently, kill prey without the aid of weapons or carry out any of the everyday survival tasks that other species undertake.

East Africa is regarded as the site where man first appeared in anything resembling his present form. Our ancestors were shorter than we are today and probably even slower at running. According to the accepted theory of evolution, early man had few natural advantages in the survival stakes, so he had to develop some special skills to compete. We know that many forms of early humans appeared and eventually died out, but the successful subspecies that was to colonize the earth proved to be the one with the most efficient brain.

We have no way of knowing what triggered the initial blooming of human brain power, but there are some clues that point the way. Early hominids had short, stumpy fingers that were incapable of fine manipulation, but they had one huge advantage and that was a natural upright stance which left their hands free for collecting food and carrying large objects, while other primates walked on all four legs. One theory for the appearance of an upright posture is that our ancestors adopted this unique carriage as a way of keeping cool in the burning tropical heat. All quadrupeds present their heads, necks, shoulders and back to the sky and absorb a huge amount of heat when the sun shines. Many take to the shadows and coolness of tree cover for the hours around noon, just to avoid overheating, but this cuts down their available feeding time. It is possible that our distant African ancestors first stood up so that less of their body surface was directly exposed to sunlight. Compared to a four-footed stance, our upright posture cuts down heat absorption by 60 per cent, allowing man to venture out when the other early primates were forced to stay in the shade. This would also explain why we still have hairy heads, as hair insulates against the heat as well as the cold. In those distant days before sharp stones were used as tools, hair would have grown throughout the life of humans and the resulting thick mat would have kept off the worst of the midday sun from their head, neck and shoulders. Standing upright would also have exposed their bodies to more breeze, increasing the cooling effect.

A very different idea of the route we took to reach our current status is based on the possibility that pre-humans went through some sort of aquatic phase in their evolution. There is a major gap in fossil evidence that spans the time between 7 million and 3 million years ago. This is an unimaginably long period and evolution must have gone through many twists and turns before ape-like man appeared. Some scientists think that instead of leaving the forests and taking to more open land, our ancestors took to the water. This could have been a way of avoiding competition with the ancestors of the great apes.

The evidence for this hypothesis is too strong to be completely ignored. There are many aspects of human anatomy that are significantly different from that of the apes and a water-based evolutionary phase would account for some of them. Few people really think that we went through a totally aquatic existence such as that of whales. Most subscribers to the idea believe that our ancestors had a semi-aquatic lifestyle similar to seals, where hunting took place in the water but breeding and resting happened on land.

This would certainly help to explain our unique absence of body hair, for thick ape-like fur would be a hindrance under water. Humans possess a layer of subcutaneous fat which, in distribution, bears a strong resemblance to whale blubber but is absent in all other primates. For a species that is reputed to have evolved in the tropics, humans are appallingly inefficient at dealing with high temperatures. Unlike other land animals that live in hot environments, we have no way of conserving moisture, in fact, we waste it in extraordinary quantities. Humans drink and sweat more than any other animal of our size, our faeces are moist and our urine highly diluted. We will die very quickly without water, even in conditions that are less than hot. In common with marine mammals, we weep copious saline tears – a mechanism that is used to reduce the salt level in a body that is constantly immersed in sea water.

When compared to apes, one of the most distinctive features of a human face is the strange nose. Why do we have nostrils situated beneath a strange fleshy hood, while apes have simple flat apertures? The sense of smell in apes and man appears to be very similar, so there is no real difference in function. It could be that our peculiar noses once acted as an anti-flooding device which prevented water entering our breathing system while swimming. It is interesting to remember that all the other apes, with the exception of bonobos, loathe water and will avoid entering it at all costs, while humans regard playing in water as a pleasurable experience.

Humans exhibit a classic dive reflex from birth. Put a baby's head into water and its heart rate immediately slows. This effect reduces the amount of oxygen used and has been adopted by marine mammals when they want to stay submerged for long periods. New-born babies also exhibit an astonishing swimming skill, at a time when they are virtually incapable of any voluntary muscle control beyond crying. Their eyes remain open under water and they swim not with chaotic splashing but with rhythmical, balanced actions; they even control the direction of movement. Based on current knowledge, it is entirely possible that our ancestors did indeed spend some time as semi-aquatic animals, hunting for fish, crustaceans and other food in the shallow waters of tropical Africa. The actions of prying and grasping in these slippery conditions would be one plausible cause for the development of our slender and highly manipulative hands. Lacking the sharp teeth, beak and talons of other fishing species, our dextrous fingers would have compensated in the finding and carrying of food. This could have been the point when arms and legs evolved into different tools; flat feet would have been ideal fins for propulsion. Unfortunately, fossil evidence for this idea remains unknown and, even if it were found, anatomy alone would be unlikely to give us clues about the behaviour of these distant creatures. Human evolution may have gone through a water phase but at some point it moved back on to dry land.

For land-based hunters the advantages of the extra height given by a two-legged posture must soon have become apparent. Elevated heads can

Tool-use was probably first confined to perfecting hunting techniques.

peer over long grass and provide more efficient long-distance sight when searching for either prey or enemies. Even now many primates will stand on their hindlegs when excited or frightened. They may be able to maintain the position only for a few seconds but it gives them an excellent view of their surroundings. Our ancestors would have learned that hands not needed for walking can be adapted for other skills. The first tools were used by man around 2 million years ago; they were simple cutting stones, created by chipping off a few flakes to modify and sharpen a rock that was naturally close to the right shape.

This tool design did not significantly alter for the next million years, but it had a profound effect on the long-term success of humans. Until the first stone was sharpened and used to kill a wild animal, feeding would have been a matter of finding fruit, shoots and other plant material. This would have been supplemented by birds' eggs, insects and reptiles that were easy to catch. Red meat would have been a rarity and probably the result of a lucky find. Tools changed this. Suddenly humans could kill larger animals and their diet improved dramatically. The image many of us have of our grunting, low-browed ancestors as rampaging, carnivorous arch-hunters is very misleading. Remains of meat meals eaten long ago are easy to identify through the way the bones have been cut. Meat-eating animals such as hyenas and big cats leave very different marks on a skeleton to those made by a human armed with a stone knife. Evidence found in caves shows clearly that, even with the aid of weapons, humans had difficulty bringing down big prey.

Skeletal analysis in Africa tells us that most meals were made up of young animals or adults with physical disabilities such as arthritis or broken limbs. In other words, humans were killing animals that were not fast enough to escape. There are very few signs of meals made up of large healthy adults or powerful species such as rhino or buffalo that could easily defend themselves. The use of tools for hunting still made a considerable difference to the welfare of early humans. Their normal diet, consisting almost entirely of available vegetation, was low in protein and required a long time to gather. The introduction of animal flesh, even once a week, was a significant step forward in the history of human nourishment.

Red meat is a concentrated form of protein. Over relatively few generations, successful hunters and their families would have put on body-weight and become more muscular – which in turn helped them catch more animals. It also gave them more 'leisure time'. One good meat meal would have been the equivalent of a whole day searching for nuts and berries. An improved diet would have resulted in lower mortality and the population would slowly have started to grow. Inevitably there would then have been increased competition for food and space, so man started to increase his range. Most tropical environments are 'user friendly' – the weather is good and food abundant – which is why the area around the Equator is home to most of the world's species of wildlife.

As early humans wandered further afield, they would have encountered conditions that were less hospitable, containing fewer animals and subject to harsher winters. If anything, this would probably have accelerated the development of their brain, for our ancestors had to find ways to adapt if they weren't to die out or be forced to move back to warmer climates, where there would be a hostile reception from the locals. Humans learned to use animal skins as clothes and to make their homes in caves in order to protect themselves against the weather. Fire soon became the ultimate survival tool and a new age began. Obviously much of this is conjecture and can never be truly proved, but the ideas are widely accepted by most scientists. Some 50,000 years ago humans were found only in the tropics and warmer belts of the Old World. Northern Europe was reached just 30,000 years ago, then Siberia 20,000 years ago, the Americas 11,000 years ago and New Zealand just 1,000 years ago. In 10,000 BC the whole human population of the earth probably totalled less than 10 million; now there are more than 5,000 million and the figure is rising alarmingly.

We are adaptable, non-specialized animals with one factor in our favour: a good brain. Our ancestors survived and passed on their genes because they had the intelligence and capability to evaluate changing conditions and work out ways of adapting to them. This took the form of building shelters, making weatherproof clothes and hunting in organized groups armed with spear throwers, slings and bows. But, most important of all, man had the power to think in the abstract, consider the future and its various possibilities and then consciously plan ahead. Life is simple for species that lead a solitary life. Apart from external factors such as

Hut-building was essential if humans were to colonize cooler, temperate zones.

weather and food supply, their actions are quite random. Communication is almost totally unnecessary as they have no information to pass on and nothing to address it to. In dealings with other members of the species, only a few simple signals need to be employed. These vary between groups but the basics are courtship signals, aggression, submission, territorial rights, etc.

When animals live in tightly knit groups, the need for communication grows. In the absence of language, evolution has produced what is called the 'pecking order'. The majority of group-living animals have developed a hierarchical system where actions are controlled by a single individual, usually the biggest. This dominant animal makes many of the everyday decisions and the others comply. Rebellion is quelled by physical punishment. There are advantages in being a member of a larger group: each individual is less vulnerable to attack and, in predatory species, a pack can often hunt more efficiently than a single animal.

There are many signals used within social groups. They come in the form of individual scent, group scent, body position, facial expressions, tail position, vocal sounds, etc. But is it language? No two people will agree on what exactly constitutes language. A wolf is perfectly capable of signalling to another pack member that he owns a certain piece of meat – so keep away. But the same signal, a growl with exposed teeth, would also be used to threaten a trespassing bear or a rival male challenging for breeding rights. This is not true language; rather, it is a simple code that conveys a limited message.

Some biologists have defined language as the ability to consider a subject that is not present. A wolf will respond if his meal is being threatened by another, but no wolf could think about a meal in three weeks' time. Only humans appear to have the abstract thoughts to consider 'what if?' and 'maybe'. These may seem to be of minor importance but they do open the doors to questions of much greater consequence. The fact that we can think about future possibilities and variations gives humans the luxury of choice and adaptability. But this would have been totally useless unless the ideas could be passed on to other members of the group, so that all could benefit or act together to make the plans work. Man is a highly social animal. We appear to have evolved as such and every clue indicates that we have always lived in family groups. In order to hunt efficiently, plan for the future and, possibly, compensate for our lack of physical prowess, humans developed language.

Working out the exact timetable of human physical evolution is difficult enough due to the scarcity of hard evidence and the numerous ways in which fossils can be interpreted. Consider the plight of those attempting to work out the history of language, for here there *are* no pieces of real evidence. The mechanics of producing words – and therefore language – are housed in soft tissue that does not fossilize. We have no real way of looking at early man and assessing the complexity of the voice box.

Controlled fire-making was a watershed in man's history. It is one of the very few primitive skills that is still relevant in the modern world.

Brains do not last long after death and leave no fossil record; we can only measure the inside of skulls and work out the size and shape of the brains that were once housed within. But this gives no sense of their capabilities and the scant information can point us totally in the wrong direction. Neanderthal man had an average brain size that was almost 10 per cent *larger* than that of a modern human, although it needs to be said that they were bigger overall, making the size of their brain proportionately little different from ours. Yet Neanderthal man used crude stone tools that did not change in design for over 150,000 years; compare this to the technological changes that have taken place in our own short lifetimes.

Neanderthals showed little sign of adopting abstract behaviour such as experimenting with art. So how advanced was their language? The truth is that we really don't know when language was first used. It may be a pointless quest for, like all other major characteristics, even a basic language would have taken hundreds of generations to evolve. Early human vocal signals were probably as basic as those of the wolf. Then nouns would have appeared and gradually the vocabulary would have expanded. The transition from primitive grunts to modern debates over the creation of the universe is a quantum leap that should completely defy our imagination. But still we take it for granted. In the late twentieth century there are a little over 5,000 known languages in the world and an unknown number have disappeared into oblivion before being recorded.

In common with the other apes, humans use a wide range of facial expressions to signal emotion. Only we have more muscles and therefore more control over our faces than any other animal. Fine movements of lips, cheeks and eyebrows in any permutation are used for signalling. We can broadcast many subtle signals without uttering a word. Some are universal and appear to be genetically controlled rather than learned. Smiling is understood in every culture; it appears spontaneously shortly after birth, even when the baby is blind and has no chance to copy the action from its parents.

Our modern smile has its origins in the fear-grin that is familiar to anyone who watches chimpanzees for any length of time. Chimps pull back their lips in an unnaturally wide 'smile' when faced with extreme danger. If the fear is generated by a dominant male, the grin is recognized as a sign of absolute submission. It shows that the subordinate animal is posing no threat. From this beginning it is just a short step for a grin to become a message of friendship. Modern humans smile for many reasons. It signals contentment and relaxation, but most of all it conveys to the viewer that the smiler is not hostile. We smile at strangers to help break the ice, without realizing that this action dates back to a time in our history when humans had no verbal language at all, when our main communication was, like that of apes, carried out with actions.

Smiling can be used in moments of extreme stress as a way of placating the aggressor. Most children at some time in their lives will be ordered by

an adult to 'wipe that stupid grin off your face'. This indicates that the
signs are being misunderstood, for the grin is not showing arrogance or
mirth but submission. In the face of threats from a dominant human,
many people use the 'smile' to signal that they accept their low status and
are not challenging authority. Just as chimps do. The modern conventional smile is simply a logical extension of this, especially to strangers. It
shows that we mean no harm and have only peaceful intentions. Other
gestures are dictated by local convention. A protruding tongue in Europe
is a mild insult, while in the New Zealand Maori society it is used as part
of a greetings ritual.

Humans as we would recognize them today have existed in some form
for around 2 million years. But cultures that we can understand are surprisingly recent. Until long after the end of the last Ice Age, about 50,000
years ago, humans were hunter-gatherers living in small family groups.
Then, slowly and spontaneously at various places around the globe, man
turned into a farmer. The dawn light of the agricultural revolution fell on
humans some 10,000 years ago and it changed us for ever.

Around that time, we believe, there was a fortunate quirk of nature that
was literally the seed of our farming efficiency. Agricultural wheat, as we
now know it, did not exist then. Its immediate ancestor was just one of
many wild grasses that grew somewhere in Asia Minor. For some reason
this seems to have accidentally crossed with a form of goat grass and,
against all the odds, produced a viable fertile hybrid. This was emmer

*The smile is a strong link with our
primate ancestors, for they
communicated with gestures rather
than sounds.*

wheat, and it came with fat, heavy seed heads that were bigger than anything seen before. This was to be the staple crop of proto-farmers. Nowadays we constantly experiment with cross-pollination and genetic engineering in order to increase crop yield, but emmer wheat was the result of a natural event.

At some stage in our history someone must have posed the question, 'What will happen if I put these seeds in the soil?' It would never have been quite as simple as that, but before agriculture really started the experimental farmers must have taken the decision to act today for the benefit of a future time that was beyond their ability to measure. This was not a programmed reflex action, but a considered, intelligent gamble. Some animals do appear to make provisions for the future. Squirrels bury nuts in the autumn and retrieve them in the cold winter months when food is scarce. But this is an instinctive response to an abundance of food, it is not a planned, cohesive strategy for survival. Squirrels born just six months earlier have no way of knowing in autumn that soon the leaves will fall and thick snow will blanket the land. They bury nuts because instinct tells them to, not their brain. Research shows that squirrels do not remember where they left the food, they find it through a combination of sporadic digging and an acute sense of smell. Sometimes they find nuts they themselves buried, sometimes they find nuts cached by other squirrels, but even more often the nuts are not found at all. Several animals show this type of behaviour, particularly rodents, but it does not signify an awareness of the future; rather, it is simply a deeply rooted response.

Human evolutionary and cultural history is peppered with watersheds that shaped our future. Relative importance lies in the eye of the beholder, but few would disagree that the discovery of controlled farming was a true milestone. Along with fire and language, agriculture was a skill that, once mastered, would change our existence. For such an earth-shattering discovery, farming took a surprisingly long time to reach the rest of the world. From its original appearance in the Near East, the new agricultural skills spread painfully slowly. One well-researched figure estimates that agriculture moved north and west at the rate of 1,000yd (1km) a year. Not all environments are suitable for sustained agriculture and the concept completely bypassed cultures in areas such as deserts which were incapable of being cultivated.

For the first time, humans had a degree of permanence. Instead of moving regularly to follow migrating animals or escape the weather, they stayed on one site. Animals were domesticated around the same time. Cattle and sheep were kept for meat and milk that supplemented the unpredictable harvests that must have been experienced by farmers who were learning as they went along. Until that point we were semi-nomads, our worldly goods confined to whatever we could carry. There was little point in developing an arsenal of weapons and tools if they were not

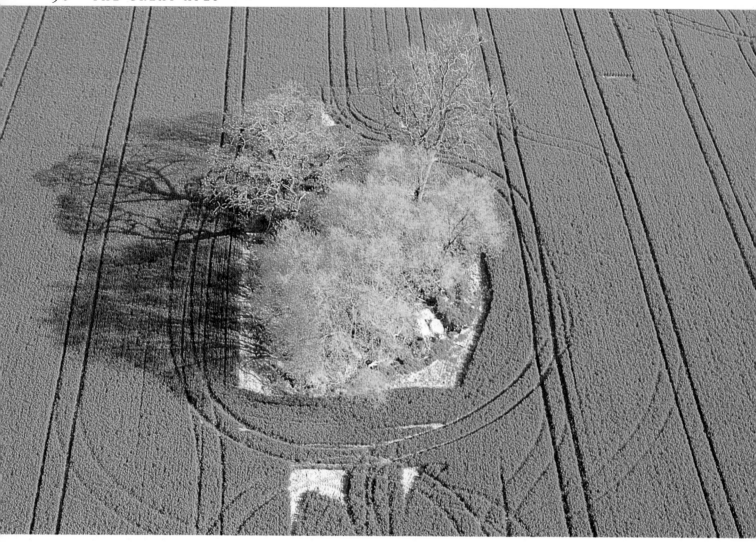

Basic agricultural skills provided early humans with the possibility of storing excess food for the future.

portable. Agricultural stability meant that buildings could be erected and large tools made – and kept – to help till the soil. Big pits could be dug where food was kept cool and stored in preparation for the cold desolation and hunger of winter. The very nature of man was changing. There is no argument about the intrinsic importance of agriculture to the success of human beings, but we need to realize that the spin-offs were not always beneficial. As agriculture permeated and eventually took over our methods of food collection, humans lost the wisdom built up by their hunter-gatherer ancestors. It created an absolute reliance on artificial methods and, if things went wrong, this could be catastrophic.

In a truly wild environment there are natural cycles that drastically reduce food supplies. Specific diseases of prey animals or food plants and drought can bring about starvation in any species population. But it is always the specialist feeders that face the greatest risk, for if their one food supply vanishes they cannot shift their attention to a different source. Our ancestors survived by eating shoots, grubs and other organic substances that we no longer see as edible. The possible results of total reliance can be

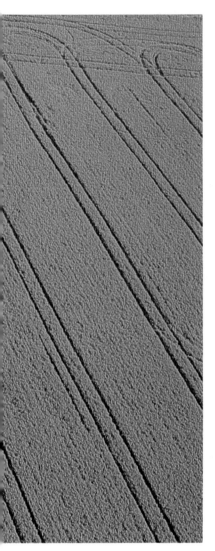

seen in the Irish Potato Famine of the nineteenth century, when blight hit the staple crop and virtually wiped it out. Hundreds of thousands died or were forced to emigrate, because the potato plants did not yield their usual harvest. Their ancestors must have found plenty to eat in the same habitat, but agriculture had swept away the early skills and few now had the knowledge to live off the land for any length of time.

The success of farming brought with it another hidden danger. Instead of having to find food by digging and scavenging, suddenly it could be grown on the doorstep through sowing, irrigating and harvesting. So more and more was planted. Eventually the chosen crop formed most, if not all, of the diet. Despite the fact that they had to work hard for it, with their wide variety of foods, pre-agricultural humans had a balanced diet containing vitamins, minerals, roughage and the occasional concentrated protein. Arable farming, however, was based almost entirely on starch-based crops such as corn (maize), rice and potatoes. Good as they are, when these substances completely take over, the diet is nutritionally poorer. Going back to skeletal analysis, we know that many early hunter-gatherer societies contained a high number of individuals who were physically bigger, had fewer bone diseases, kept more of their teeth and often lived longer than people in later societies who relied on a monoculture farming principle. All because their diet was better. To quote just one example, late Ice Age men in what is now Greece had an average height of 1.78m (5ft 10in); 2,000 years after the adoption of agriculture, this had fallen to 1.63m (5ft 4in). A similar percentage drop also applied to women.

It is ironic that the early man-creatures roaming East Africa would probably have had a diet similar to today's chimpanzees but, with the triumph of agriculture, the quality of human nourishment subsequently plummeted. Due entirely to our intelligence and creativity, for thousands of years the human diet was poorer than that of the apes, most of whom consume several hundred distinct sorts of food. In the West today we rely on less than twenty plant species for 95 per cent of our basic non-meat diet. Whatever the long-term results of farming, the system has one undeniable advantage: cultivated land always produces more edible material than a wild patch covering the same area. This means that although the menu may be less than exciting, more people can eat, and that has always been the one overriding consideration.

Farming marked the first of several revolutions that were to radically influence every aspect of our existence. The earliest known written language appeared about 5,000 years ago, with a complex pattern of syntax and grammar. The first known wheel was used about the same period, but even this hugely important invention has not reached every corner of the world. There are still cultures that have not adopted this most basic of human gadgets – and see no reason to.

There have been several milestone 'revolutions' since agriculture. Industrial, electrical and atomic have all had their effect on our lives. Our history spans around 2 million years, yet most technological advances have taken place in the last century. At first glance this is an unbalanced development, but it should not be that surprising. Scientists working now are building on research carried out by their predecessors over centuries; they are plugging into a vast network of accumulated knowledge. Inventors now do not have to work out a power source before constructing a computer; they simply plug in and throw a switch, leaving more time for their own specialities. The basic components of high-tech inventions can be bought off the shelf, instead of being built from scratch. Technical education for children was a rarity until the early twentieth century; now most people leave school with at least a rough idea of the principles that govern science. And, finally, there are more people today working in science than in the rest of history put together. The reservoir of resources and knowledge alone explains why there has been a mushroom effect in scientific development.

We may live in a high-tech society but our automatic actions are still guided by basic instincts shared by the other anthropoid apes. We are starting to realize that even humans use body language unconsciously to express emotions. Due to the high mobility of people, the telephone is now the most important means of instant communication. It is well worth watching someone talking on the phone as they try to put over an important point, or, better still, become angry. Arms are waved about, fingers pointed and faces fold in scowls. To no avail, for the recipient on the other end of the line can see none of it. This is purely instinctive, just as with the apes. We use body language without thinking about it. It is estimated that humans use around 3,000 different hand signals as part of their non-vocal communication. Like languages and dialects, these are different throughout the world and no one culture uses all of them.

Gestures need to be seen in context to be fully understood. Gorillas often mate face to face and usually keep close eye-to-eye contact throughout. This helps to maintain the bond, making the act personal. In any relationship between close individual humans, unbroken eye contact is also important; it can be seen between mother and child, lovers and close friends. But without the underlying security of this relationship, the stare takes on a different meaning. When rival silverback gorillas meet, they glare at each other. This is a direct challenge, as neither will look away and accept the dominance of the other. It builds up the tension and frequently results in more aggressive displays or even violence. Similar scenes take place when a stranger walks into a bar and stares directly at a local man. If the eye contact is prolonged and unbroken, the atmosphere thickens. We are not comfortable in this situation, for no reason other than that it is a threat-relic of our man-ape past. The feeling of unease is effectively used by people in dominant positions; they frequently use an unblinking gaze to 'stare down' subordinates, making them more submissive. This can clearly be seen in boxers, teachers and even parents.

Humans evolved as social animals, so had to learn to live peaceably within a settlement, and this is not an easy matter. Every group is made up of individuals, each with their own temperament and needs, yet they have to act cohesively to survive. One of the ways animals achieve this is by copying the actions of a single individual or the dominant core. Gorilla groups are guided by the silverback. When he sleeps, everyone sleeps; when he starts to build a nest, they all do. Overt compliance is an excellent method of avoiding conflict. Humans are not quite so straightforward. When subconsciously giving out signals, we often use postural echo – or, in other words, we copy each other. If two friends sit together talking and one crosses his legs, it is more than likely that the other will soon do the same. If one leans forward and rests his hands on his knees, it will not be long before the other copies the movement. This kind of group behaviour can be seen by any interested observer yet is rarely noticed by the parties involved. This is exactly what apes do in the same circumstances in order to pass friendly signals to each other and prevent friction within the group. Resting chimpanzees spend a great deal of time delicately combing through the fur of neighbouring animals. Mutual preening of this nature has little to do with keeping clean; its prime function is to strengthen bonds by emphasizing group dependency. Humans display exactly the same behaviour all the time. How often do we see a wife removing a fleck of dust from her husband's coat before they leave the house. Or a father straighten his young son's tie as he sets off for school. We all do it automatically, not for cosmetic reasons but for physical contact and mutual reassurance.

Whenever animals live in structured societies, there is likely to be competition for food, mates and territory from neighbouring groups of the same species. It is impossible to generalize about the nature of such rivalry, for there are species, such as badgers, where individuals freely travel between groups and aggression levels are low. But in most cases, competition is so fierce that outright war ensues when rival bands meet. The whole group may join in the battle or it might be only the dominant males that fight. Humans are not equipped to survive alone. Our ancestors, until comparatively recently, would have lived in extended family groups made up of less than a few dozen people. Then almost everyone was related and all members knew each other well. Occupants of nearby villages may have been known and recognized, but they would also have been treated with suspicion. Twentieth-century life has changed this culture to an incredible degree.

Most humans are now city dwellers, surrounded by literally thousands of total strangers. Psychologists tell us that few people are comfortable with this, so we re-create small-scale, more comprehensible, close-knit communities by joining golf clubs or supporting local football teams. Other clubs and teams then become the token 'enemy', to be mocked and defeated whenever possible. Membership of the group is advertised with scarves, badges and songs, each of which helps maintain the internal bond between individuals. At a primitive level the superficial interest that adheres the membership is probably less important than being a member of a manageable, friendly 'family'. Be they football supporters or a rival quiz team, the opposition are simply outsiders. This is what prompts the majority of sports-related violence. Soccer hooligans are lapsing into behaviour that their ancestors would have adopted when faced with another band of hunter-gatherers. Humans live in a sophisticated world, but our behaviour has not caught up and is still remarkably close to that of the fur-clad hunters that once lived in caves. Each of us has a basic desire to be a member of a small group in order to make sense of social structure and remain an individual. When ape groups become too large, one or more wander off to start another – which is how our ancestors would have responded to a similar situation. Humans cannot cope with the rapid changes that have occurred in our society. Instead of living in small groups, the majority of humans are now brought up in tightly packed colonies containing millions of individuals. So we should not be dumbfounded to find that urban man suffers far more from stress-related problems than any tribesman living in a small family group in New Guinea or Brazil.

Humans are a phenomenally successful animal. We are the only species in history that has the power to dictate not only our own fate but also that of other species that share the earth with us. Many believe now that we are too successful. Modern medical science has either eliminated or enabled us to control most of the diseases that once kept our population in check. In the Middle Ages few people died of old age and those that did were often buried before they reached fifty. An increasingly high percentage of us can realistically expect to see our eightieth birthday – everyone is living longer.

Frighteningly lethal hunting methods have wiped out the majority of serious predators that threatened our survival and agriculturalists are rapidly learning to coax crops from areas that were once barren scrubland. We have conquered most of the natural factors that normally restrict the population growth of every other animal. In any stable environment, the numbers of each resident species is limited by food, predation, disease and living space. Without these limitations the human population is doubling approximately every thirty-five years. It is mathematically provable that the earth cannot sustain this unimaginable growth rate.

The reign of humans on earth has been directly compared to fleas infecting a dog. This may be insulting but the analogy is accurate. Parasites can exist only on a host animal that supplies food, warmth and shelter. From a strictly biological viewpoint, it is in the fleas' interest to expand their population, so they breed. It is equally in their interest that the dog stays fit and healthy, but occasionally the number of fleas grows too high and they take too much blood when feeding. The dog then becomes anaemic and dies from a secondary infection. With the loss of their host, the fleas also die out, for they have nowhere to live and nothing to eat. If the human population continues to expand, in 200 years we will have exhausted the reserves of oil and other resources that keep twentieth-century humans alive and mobile. The air will have become too polluted to breathe and our waste will contaminate our drinking water. On top of this many 'wild' habitats will have been turned over to crop culture and 70 per cent of the world's wildlife will have died out.

At the moment we calculate that there are around 30,000,000 species of organic life on earth. Realistic estimates indicate that around half of these will disappear by the year 2100. That is a wholesale extinction at the rate of seventeen species an hour. Mankind has lost touch with the idea that we are merely part of the natural world rather than masters of it, even though our brain gives us the ability to control our surroundings. While humans were confined to tiny isolated populations in the tropics, our actions were ecologically irrelevant. Now they have a major impact on the entire globe. We have just two choices in the very near future: we must learn to consume less and drop our breeding rate or attempt to live with the consequences of ignoring the problems. Our ultimate nemesis is unlikely to appear in the form of universal destruction on a biblical scale. Total extinction of such a successful species is difficult to imagine, although it did happen to the dinosaurs. More probable would be a new disease that spread rapidly as a result of high population density, lowered physical strength through a worsening diet and unclean water. There are many precedents for this scenario. When European rabbits contracted myxomatosis, 98 per cent of them died within six months. The same mortality is entirely possible in humans, should a similar virus infect urban areas where thousands of us are packed into a few square kilometres. The city habitat, with its high population of constantly moving and mingling people, all living in warm buildings, is a perfect breeding and transmission medium for a huge number of known diseases.

Unlike bigger animals, disease-creating 'bugs' are constantly mutating, and because their life cycles are so short, their evolution is rapid. The common cold is a perfect example. Just as we build up an immunity to one form, it mutates into something slightly different that cannot be recognized or dispatched by our antibodies. There is a widespread theory that our final fate doesn't lie in the fall-out of atomic warfare but in decimation by a virus already known as harmless that, in some altered form, will break down a critical function of our body system.

This may all sound rather bleak and obviously the future could be very different, but few biologists would disagree with the hypothesis. Man is now too successful for his own good we are in danger of engulfing the planet that we need for survival.

6
FOOD:
What Do Apes
Really Eat?

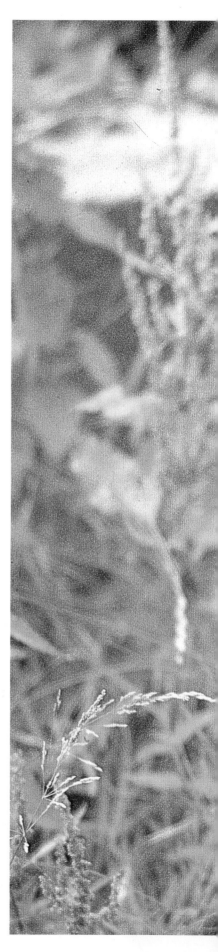

IMAGES IN THE POPULAR media seem to suggest that apes eat nothing but bananas. In fact, most wild apes never see a banana in their lives and their diet actually consists of several hundred different foods. Each of the great apes is basically vegetarian; among other things they will eat shoots, leaves, fruit, ferns, flowers, fungi and cultivated crops, but then they will also eat birds' eggs, insects and red meat.

Defining the diet of a chimpanzee or gorilla is almost as hard as drawing up a typical human diet. We eat whatever is palatable and available, and this is tempered by cultural and seasonal factors. Apes' food is controlled to a great extent by their geographical location and the time of year, along with their sex and age. Personal preferences also play an important part.

To the uninitiated, a tropical forest might be thought of as a rich woodland stuffed full of animals, with fruit and nuts dripping from every tree. It is true that at certain times some trees might be laden with fruit, but not all trees produce seed in this form; also some fruit is not edible and prized fruit is quickly taken by birds, bats and arboreal mammals. In other words, locating food in a rain forest is not quite as easy as it might seem.

Eating, then, is a time-consuming task that involves travelling to reach trees at just the right moment. And even then the soft inner fruit may be protected by a thick skin or layer of spines that needs to be stripped off to reveal its contents. Finding food is the main reason why apes move around within their territories. The group-living chimpanzees and gorillas need a large amount of nutrition to support every member of the community and they would quickly exhaust the possibilities if they stayed in one area too long. A single adult gorilla can eat up to 30kg (66lb) of bulk in a single day. This would be an unsustainable drain if the apes fed in one small place, so they are forced to move on regularly. Constant wandering gives them a subtly changing diet. As they move on to new altitudes or forest

Bonobos eat a wide range of vegetation. Their large and flat teeth are perfectly designed for chewing tough plant material.

areas with different soil types there is a shift in the type of vegetation, providing a broad-based variety of foods. It also gives a chance for the exploited feeding area to recover.

Apes spend much of every day feeding because their food can be surprisingly low in nutrients; what it lacks in quality the apes make up for in quantity. Visitors to zoos are accustomed to seeing fat or even obese apes, particularly gorillas and orang-utans. This is partly due to a chronic lack of exercise, but also because their diet is often too rich, giving them far more calories than they would ever ingest under natural conditions. Obesity is almost unknown in wild ape populations, although they do have a round-bellied appearance because so much of their diet is made up of leaves and shoots. This tough, fibrous food is slow to digest, so, like a great many other leaf-eating animals, apes eat a lot and keep the food in their stomachs for a long time in order to process it thoroughly. This produces a huge, vat-like abdomen consisting of the current meal rather than excess fat. The feeding strategies of each species are so different, it will be easier to deal with them separately.

COMMON CHIMPANZEES

Chimps have a far more diverse diet than the other apes and much of their menu is dictated by 'cultural' differences. Apes learn from their parents and peers. In some parts of the chimpanzees' range they developed eating habits that are unique. In western Tanzania there are two prime study areas for chimpanzees, Mahale and Gombe Stream. Here there are 286

Young apes will take milk for as long as they are allowed.

potential different foods that are common to both sites but of these only 104 are eaten by both groups. The remainder are taken by just one of the populations. There appears to be no logical explanation for this curious anomaly. The only conclusion we can draw is that each group has learned what is edible and they ignore everything else. Because they have each learned different lessons, both groups are missing out on potentially worthwhile food supplies.

These social quirks have led to some strange dietary differences. Oil-nut palms are widely distributed across the range of common chimps. Some chimps eat only the pith, others take the kernel; at Mahale every part of this tree is ignored completely, while at Gombe it provides the most important food item for the resident chimps.

A single group of chimpanzees might eat more than 200 species of plant and up to thirty species of insects. An 'average' breakdown of their diet would be 65 per cent fruit, 30 per cent leaves and shoots and 5 per cent animal matter. A list like this can be very misleading, for it changes from day to day and from chimp to chimp even within a tiny population. To complicate matters even further, chimps have 'food fads' that are surprisingly similar to those of a human. Individuals go through periods when they enthusiastically eat one particular food at every opportunity; a week later they will walk past it without a second glance. Hungry chimps will eat anything remotely edible; when food is plentiful they choose only the most tempting treats and leave everything else. Given the opportunity, chimpanzees will eat as much as possible in a single sitting. This gives them more time to rest and groom.

There is one aspect of the chimps' diet that often surprises even keen wildlife enthusiasts and that is the fact that they will eat red meat. Not only do they eat it; they will kill to get it. In recent years zoologists were taken aback when they discovered just how little was really appreciated about the diet of wild chimpanzees. It was always assumed that they were exclusively vegetarian and occasionally supplemented their menu with the odd insect. This was completely wrong. Chimpanzees are skilled hunters of large mammals and they feed off flesh with a relish that would do justice to a true carnivore.

Common chimpanzees, in the right frame of mind, will take meat in any form they can find it. They are opportunistic hunters and will grab at any small animal that comes within reach. Mice, rats and young birds are often caught and quickly eaten. Chimps are not above a spot of robbery by intimidation; they are particularly keen on stealing meat from baboons, no mean hunters themselves, through size and sheer weight of numbers. But these are just snacking tactics. When they really mean business chimpanzees band together in what can only be described as an organized hunting party. Monkeys and bush-pigs are the prime targets but no small mammal is truly safe when a chimp hunting pack is looking for meat.

Males do most of the killing of big animals but, in a large-scale hunt,

Chimpanzees often collect food and carry it away to eat.

females and adolescents can act as beaters to drive the prey into the trap. Males will often hunt alone or in pairs, but a co-ordinated group hunt can last over an hour and has the hallmarks of a well-practised strategy. Generally the selected prey is driven towards a site occupied by the biggest, fastest males. The beaters shriek and make as much noise as possible to frighten and disorientate the prey. But no two hunts are the same, for chimps have the versatility to respond to changes in circumstance. They will block off likely escape routes or completely surround the chosen animal. If it stops and takes shelter in a tree, a chimp climbs up to keep it moving towards the trap. One remarkable feature of a mass hunt is that the normally volatile and noisy male chimpanzees maintain complete silence as they stalk their prey.

The end generally comes quickly. Big animals – particularly monkeys – are simply pulled apart. Smaller ones are beaten against the ground or a branch with powerful overarm swings. Chimps sometimes kill with bites through the spinal cord or by wringing the necks of their victims. Truly unfortunate animals may be eaten before they have been killed. At the moment of capture, there is enormous excitement throughout the hunting party; the chimps howl and shake branches, making the woodland echo with their celebration. Big males get first choice of the pickings and they do not readily relinquish their prize. Lower-status animals always attempt to beg or even steal pieces of meat, but they are allowed to feed only when the males have had their fill. Chimpanzees do not bolt their food; they eat it slowly and often chew pieces of meat together with a small leaf. The brain appears to be the real treat of the hunt and it is the most eagerly sought-after part of the prey. Chimps need to work hard to reach the soft interior of a mammal skull. The mere fact that they spend so long breaking and biting off pieces of thick bone is enough to tell us that they feel the effort is worth the reward. The chimps' appetite for meat is so strong that younger animals can be seen gnawing on completely stripped bones several days after the kill.

Hunting is not a regular or frequent event. A chimp group can go many weeks without attempting to eat meat and then suddenly it will stage two hunting raids within three days. They often prey on local young baboons, a species they play with in their infant days and then feed alongside throughout the rest of their lives. There are at least three well-documented cases where human babies have been killed and eaten by chimps – something hard to imagine when watching these animals dressed up in shorts and sweaters drinking tea and eating delicate sandwiches. Zoo chimps are just as likely to hunt as their wild counterparts given the opportunity. One keeper in an English zoo watched helplessly as a stray cat jumped off a wall into a large pen housing eight chimpanzees. The hunt began immediately and in sixty seconds the cat was caught by an adult male who had probably never killed before in his life; the victim disappeared beneath a scrum of hooting chimps. The following day just a few large bones and fur scraps were found.

Cannibalism has been seen in enough study groups to suggest that it is far from rare. Young chimps are killed and eaten as part of raids against rival groups, but some animals will kill inside their own groups. It was first recorded in 1971 when a scientist watched as a live, partially eaten new-born chimp was passed around a group of adult males, who all joined in the meal. Males are the usual culprits of cannibalism within their own group but at Gombe Stream one female, known as Pom, went through several phases of killing and eating infants.

Bonobos are not thought to share the common chimpanzees' passion for red meat. Their intake of animal protein comes from invertebrates rather than larger animals. They eat the odd insect as they come across them, but they certainly don't expend the same energy and expertise in the search as common chimps. They have not been seen indulging in cannibalism, although this could be the result of the relatively short time spent observing them. After all, we once thought that chimps were vegetarian.

For details on the use of tools in feeding see Chapter 7.

ORANG-UTANS

The one outstanding feature of an orang-utan's feeding strategy is the detailed memory that each animal must employ in order to keep track of all the sites regularly used. Throughout the distribution of the species more than 400 orang-utan food plants are recorded. Within the range of just one animal, there may be almost half this number. Orang-utans are slow-moving animals that cover a huge area of forest in a year. They can't quickly flit from site to site eating as they go because they aren't nimble enough for that sort of fast-food culture. They are forced to be more methodical and thoughtful in their travels.

One of their favourite foods is the fruit of the durian tree. These look a little like thorny melons and, certainly from a human's viewpoint, offer a sweet-tasting but foul-smelling meal, as they are blessed with an odour strongly reminiscent of rotten eggs. But orang-utans adore them. Unfortunately, the durian fruits around once every two years and the trees are often well spaced throughout the forest. Orang-utans would miss out on the harvest if they relied purely on luck to find the tree at just the right moment and there is overwhelming evidence to show that the durian feasts are not a matter of good fortune. Orang-utans appear to know exactly where to find each of the durian trees within their range and, as if that wasn't impressive enough, they are equipped with an accurate internal calendar that keeps track of the fruiting season of each tree.

Unlike European deciduous forests, where the fruiting season is short and extremely predictable, tropical rain forests are disorganized in their fruit harvest. Not only does each species of tree fruit at different times of the year but also each individual tree has its own cycle. Speaking personally, I would find it impossible to pinpoint a single week over a two-year period without the help of a diary. Yet orang-utans can, and not just for one tree but for the several dozen that might be found in their feeding area.

Around a third of the orang-utans' diet consists of young leaves and shoots, along with occasional insects, eggs and even young birds. Orang-utans have powerful jaws and strong teeth for opening nuts and hard-shelled fruits. They have even been seen munching their way through tree bark. Like many animals, they also eat mineral-rich soil. Around 65 per cent of an orang-utan's diet is made up of fruit, and the apes themselves are directly responsible for the propagation of the trees that provide the bulk of their food. Seeds are swallowed at the feeding site and slowly pass through the orang-utan's digestive system. When they are finally dropped the seeds have been thoroughly softened by their passage through the orang-utan and emerge encased in a rich manure which will certainly help their chances of germination. By the time the seeds reappear, the ape is likely to have moved away from the mother tree and any resulting new

Young orang-utans feed on easily obtained leaves for a while after weaning.

A rich crop of fallen fruits is one of the few attractions that tempt female orang-utans on to the forest floor.

plants will not directly compete for growing space with either the original tree or a host of other saplings packed tightly together, having fallen from the same source. This is an elegant system that randomly scatters trees through the forest and encourages the growth of a constant new food supply for the benefit of future generations.

Strange as it might seem, finding water can be a real problem in rain forests. The soft, peaty woodland floor absorbs water in seconds and rivers may be several miles away. Thirsty orang-utans will drink rainwater caught in suitably cup-shaped leaves or hollow tree stumps, but this supply quickly becomes contaminated in the tropics and must taste less than refreshing. Orang-utans have learned to exploit a much better source. Immediately after rain, they spend some time carefully licking their arms and hands to take in the fresh droplets of water. They will also dip their forearms into streams and rivers, then carefully suck the hairs dry. I watched one young baby who hadn't mastered the technique of licking her own hair; after heavy rain she simply sucked the water off her mother.

GORILLAS

Gorillas spend over half their waking time feeding. This allows them the opportunity to sample a wide range of food plants in every session. Multiply this by the number of days in a gorilla's life and then consider the number of plant species that grow in a tropical forest, without forgetting that gorillas may roam widely and are free to exploit many different habitats. For these reasons, it is impossible to draw up an exclusive list of gorilla food items.

Gorillas have been known to eat meat, but it is confined to very small prey. Arthropods, a huge invertebrate group made up of insects, millepedes, etc., will be casually eaten by any gorilla that happens to find one during a normal feeding session, and individuals do develop a real taste for insects and deliberately search for them. One survey showed over 100 different food plants were eaten by a single gorilla group, and the true total is much higher than this. Bananas, several vines and ferns make up a large chunk of the gorillas' food intake. Much of their food is found in secondary forest, where the original trees have been felled and new growth has appeared. Often these areas have a wider variety of plants at ground level and they also offer more young shoots of all species. It is these tender delicacies that tempt gorillas to feed on roadside verges and on the edge of agricultural land. This opportunism has accustomed some gorillas to the presence of man and a small number make the transition from foragers to crop-raiders, incurring the anger and retribution of the farmers. Gorillas, like many other animals, have some inbuilt mechanism that triggers off an urge to replenish minerals when levels fall in their body. There is a site on Mount Mikeno, in Zaire, where mountain gorillas have been known to chew and swallow a volcanic soil that is very rich in potassium and sodium.

Plant material is low in nutrition, so adult gorillas need to eat huge amounts every day.

Gorillas use their hands as the main tool for finding and preparing food and they have different ways of handling plants. Bamboo shoots are dug up and peeled before being chewed. Stems of wild celery, a popular food with gorillas but one I found too bitter to eat, are opened up and only the soft centre is eaten. George Schaller, who carried out some detailed work on the dietary content of the food plants, showed that the celery leaves thrown away by the gorillas have a higher nutritional value than the plant parts that are actually eaten. This might be a culturally learned action or could simply show that the gorillas prefer the stem centres.

In common with all apes, gorillas in captivity almost always pick bananas first when presented with a pile of assorted food. But in the wild fruit makes up a small percentage of their daily intake. Eastern lowland and mountain gorillas eat much less fruit than the western lowland animals, but this is through a lack of opportunity rather than any preference they might have: fruit trees are just more numerous in the range of western gorillas. Fruit-eating gorillas live in smaller groups than either of their cousins and that could be an effect of their diet. Edible leaves are easily picked and usually grow in huge quantities, so mountain gorillas do not have to work hard or travel far to find a meal. On the other hand, fruit is seasonal and appears in much smaller quantities. The limited supply of fruiting trees within a foraging range may be a factor in controlling the number of gorillas in a single group. A large number of animals would probably not find enough food inside the area covered in a day. Obviously they don't eat just fruit; their diet includes many other parts of plants. We might speculate that they enjoy fruit so much they are not willing to share the limited supply among members of a larger group.

Despite their bulk, gorillas can be very delicate when feeding. They often peel away the outside layers of stems to reach the soft core.

7
BRAIN POWER

FOR ZOOLOGISTS and behavioural scientists intelligence is the most elusive of all animal characteristics to measure. No one has yet formulated a universally acceptable definition for human intelligence, let alone for that of wild animals. Although the physical size of an animal's brain obviously has some bearing on its function, we know now that a heavy brain does not mean a giant intellect. Darwin himself wrote: 'One cannot measure intelligence in cubic centimetres.' Brain power is not a quantifiable commodity, simply because we do not understand exactly what it is.

Humans are seen as intelligent because, among other things, we are highly successful, have colonized the planet and have a growing population. The same can be said about slugs, and yet they have no brain in the form that we recognize. Success, then, is no yardstick of brain power. Memory has often been linked to intelligence: after all, it does take a great deal of 'brain' capacity to retain reams of information in the long term. But personal computers can store staggering amounts of data, even though, currently, their creative thinking abilities are completely non-existent.

Does complex language indicate a sophisticated brain? The diverse communication skills of humans are certainly among the major factors that make us what we are. But language is not confined to verbal sounds. Bees have a wonderfully efficient way of telling other members of their hive the whereabouts of a rich food source. Using body movements they can signal the direction of the food in relation to the sun, and how far away it lies. This is amazingly complicated information that could never be understood and passed on by creatures with far more developed brains. Yet bees could hardly be regarded as highly intelligent. We need to remember that language is important only to gregarious species that have a need to communicate between themselves in order to survive. Animals with solitary lifestyles would have little use for complex communication, no matter how intellectually advanced they might be.

Today's accepted wisdom sees true intelligence lying somewhere within the realms of adaptive skills. Theoretically, how an organism evaluates and responds to an unknown set of conditions should be a fair guide to its cerebral power. Unfortunately, this does not work for every animal. Some are genetically programmed to thoroughly investigate everything new, and have the skills to take full advantage of most opportunities. The

Young gorillas show a skill for language that would have surprised early researchers.

house mouse is a perfect example. This is one of the world's most successful species, occupying every continent and almost every conceivable habitat. They have the ability to fully utilize the world around them, but do they consciously evaluate it? Devious experiments have been dreamed up to show that mice quickly learn to negotiate convoluted mazes in order to reach food. They can be taught to press yellow buttons on machines to be rewarded with grains of seed. Many animals never master even this simple technique and would starve to death before learning to press the right button. At this level house mice certainly appear to be bright enough, but they have not adopted tool use or even simple language.

Some species are physically limited in their application of intelligence. Whales and dolphins are often nominated as strong contenders for the title of 'most intelligent' animal. Evaluating this claim is a nightmare. Their sheer size makes conventional controlled testing almost impossible. Cetaceans have evolved for life in a highly specialized environment; they have no need to adapt and explore in the same way as land mammals. Neither can they manipulate tools, building blocks or any of the other gadgets used in intelligence experiments. Yet preliminary tests show that they have an impressive capacity for learning and understanding a wide range of human sound signals. They can 'read' signs and appear to have a well-developed communication system of their own. At the moment the language of whales is a mystery to us, although scientists have speculated about its complexity and content. Some believe that it may rival human language in the abstract and subtle ideas it carries.

Intelligence obviously operates on many different levels. Humans with severe learning difficulties can display genius-level gifts in pure mathematics or music, yet these same individuals may be unable to tie their own shoelaces. Can the mastery of one skill be seen as intelligence? Or is it simply a quirk of genetics that has endowed an individual with this single extraordinary talent while leaving the rest of the brain underdeveloped. Similar arguments can be applied to entire species. An animal that learns to negotiate highly complex maze systems can only, safely, be termed a master maze navigator and nothing more. Perhaps the quest for definitive relative intelligence tables should be abandoned until we understand the question a little better. Some scientists are convinced we will never be in a position to analyse intelligence properly.

When talking of comparative intelligence, all we can say is that apes *appear* to be among the brightest of all non-human animals. And that is using a reference scale based on our own brain, which is regarded as being the ultimate. As we see ourselves as intelligent, any animal that behaves in similar ways is inevitably going to seem bright. There is something undeniably fascinating about the attempt to understand the thought processes of other animals and comparing them to our own. It is an obvious method of reinforcing our own status as *Homo sapiens* – wise beings.

Chimpanzees spend more time playing than almost any other living animal. One theory links play activity directly to intelligence, as it is seen as a mechanism for learning.

Through well-designed experiments it is possible to assess an ape's ability at handling specific tasks, but even that is open to interpretation. Problem-solving behaviour of captive animals, or wild individuals in close and regular contact with humans, can simply be the result of mimicry rather than inventive thought. Apes are genetically programmed to copy their elders and they are equally willing to mimic human behaviour under the right conditions. As to the question of the 'most intelligent' ape, answers are very subjective and varied. There is a semi-serious joke among primate keepers that tells of an experiment where a screwdriver was offered to each of the great apes in a zoo. The chimpanzee quickly looked and then threw it away; the gorilla lazily scratched its back with it; but the orang-utan hung on to it until the keeper went off duty – he then used the screwdriver to unpick the lock and escape. Zoo keepers and others who work with apes on a practical rather than a scientific level often think that orang-utans are the brightest of the three, and as they are with their animals for eight hours a day for years at a time, their opinion should not be dismissed.

The screwdriver story, like many jokes, is based on truth, because orang-utans really are the master escapologists of the ape world. Their persistence and originality are challenges to any zoo keeper. The modern practice of keeping apes in open enclosures rather than steel and concrete boxes has given them many more opportunities for showing their skills. Orang-utans use branches to ford deep moats that surround many pens. They also take out the centre pins from hinges in order to open doors the wrong way, they unravel chain-link fences and even test the current of electric fences. A male called Fu Manchu, who lived in Omaha, regularly unnerved his keepers, who would find him wandering the grounds of the zoo, having been locked up the night before. It took some time for them to realize that he had learned to pick the lock of his cage; he kept the short wire he used as a skeleton key hidden in his mouth whenever humans were near.

Institutions throughout the world have been working on ape intelligence for more than seven decades. The biologist Wolfgang Kohler carried out pioneer studies as early as 1914 at a primate research unit on the island of Tenerife in the Canaries. Since that time almost every conceivable evaluation technique has been explored, ranging from tests with totally wild animals to taking a new-born ape into a domestic house and raising it exactly as a human baby. Nearly a century on from the earliest experiments, little has been absolutely resolved.

Most early tests were based on the ape solving mechanical problems in order to collect food. Some were extremely simple. Fruit was suspended from a ceiling out of the apes' reach and empty boxes were scattered around the room. Some animals jumped and strained at the fruit, with no success, and they finally gave up. Others played randomly with the boxes until, more by luck than anything else, a tower was built and the fruit was within their grasp. Several showed immediately that they understood the problem by piling the boxes together to use as a climbing frame. The results were very mixed. Some apes built towers but in the wrong place; others had the right idea but lacked the skill to make a stable structure. When shown by another ape or even by a human, some of the less innovative animals were able to master the technique. It was soon understood that ape intelligence, like that of humans, varies from individual to individual.

We can now see that some tests were, unwittingly, heavily loaded in the favour of certain species. One researcher placed food inside a long tube, open at both ends, and different apes were given poles to help them reach the reward. Both chimpanzees and orang-utans solved the problem by pushing the food out. Gorillas tended to perform poorly at this and other manipulative experiments. Orang-utans and chimps spend a great deal of time in trees, and have evolved highly dextrous hands for both climbing and obtaining food found in awkward places. Gorillas, on the other hand, are terrestrial animals that only occasionally venture into trees. The bulk of their food is found on the ground and therefore they have less need for searching and probing fingers. This does not mean that gorillas are any less intelligent.

The difference in natural feeding techniques led early scientists to believe that chimps were the brightest of the apes, followed by orang-utans with gorillas coming a very poor third. This simply reflected the fact that tests were unintentionally geared to the chimpanzees' highly inquisitive, wide-ranging feeding strategy. In the wild they investigate any new object thoroughly to assess its potential as food. Further observations show that chimps have a shorter attention span than the others and are prone to lose interest if the reward does not come quickly. Food that is visible but beyond reach can quickly bring about a screaming, floor-slapping temper tantrum in a chimp, while the less excitable gorilla is more likely to persevere and eventually succeed. Orang-utans often adopt the direct approach when food is hidden inside tubes or boxes; they ignore the offered tool and attempt to smash the apparatus to get the fruit. Many researchers now believe that there is so much individual variation in skills and because each of the great apes shows talent at different tests, it is impossible to award the 'most intelligent' title to any species.

Following tool tests, then came the thorny question of language. This opened up a whole new problem of definition. Can an animal be said to understand a language if it passively recognizes and understands phrases, or does it also need to be able to respond in similar terms to be regarded as truly comprehending?

Projects were conducted as early as 1916, when a chimpanzee was taught to say the word 'Mama'. But merely repeating a sound does not signify the use of real language. Parrots will mimic an astonishing range of words and non-linguistic sounds, but they have no idea what they represent. In the 1930s a young chimpanzee by the name of Gua was shown to grasp the meaning of English phrases far quicker than a young human child, but the ape soon reached saturation point while the child's learning skills constantly accelerated. In 1951 Keith and Kathy Hayes taught Vicki, a chimpanzee, to say 'Papa', 'Mama', 'cup' and 'up', but it is doubtful whether the animal actually understood the symbolism; it was just repeating sounds. And even that was not a great success, for Vicki's pronunciation was clumsy and slow.

One of the early apparent successes in language acquisition by apes was seen when Beatrice and Allan Gardner took Washoe, a wild-born female chimpanzee, and housed her in a caravan in their Nevada garden. Judging by her weight and physical development, Washoe was between eight and fourteen months old when she moved in with the Gardners. She was provided with toys, tools, pictures and magazines indoors; outside she had access to trees and swings for play. It was reasoned that if Washoe was ever going to communicate, she must have interesting things to talk about and someone to say them to. Most days were spent with humans, but she was left alone at night, in exactly the same way a Western child of the same age would be.

Anatomists agree that apes simply do not have the vocal equipment to produce the complicated series of subtle sounds employed by humans to achieve spoken language. They can make a range of noises but are probably incapable of articulating and controlling the delicate but perceptible changes in vowel sounds that are so dominant in most languages. Experts in brain function tell us that apes do not have the correct speech control

Chimpanzees are very inquisitive and will investigate any new object, especially if they think it might contain a reward.

centres to drive a verbal language as we would understand it. But it would be wrong to assume that language exists only as sounds produced via the larynx.

These seemingly fundamental barriers, along with other considerations, prompted the Gardners to abandon ideas of spoken language and opt for American Sign Language (ASL). ASL is the widely used hand language of the deaf and offered a practical, thoroughly tested mechanism for two-way communication that did not rely on vocal skills. It was decided from the very beginning that all communication with Washoe, and even between humans in her presence, would be carried out in ASL. Speech was to be kept to a minimum whenever possible. To avoid the possibility of Washoe recognizing subliminal body-language, which she would soon learn to decipher from familiar humans and could incorporate into her understanding of situations, she was exposed to many different people each conversing in ASL. This meant that the researchers had first to master the signing code. It was a major project.

Washoe was treated like an infant human from the beginning. Just as parents repeat words for familiar objects such as toys and food, Washoe was shown the relevant signs for everyday items. Teaching took the form of play rather than intensive tuition sessions. It was hoped that this approach would mimic the normal conditions experienced by wild chimps and therefore hold her attention for longer. At first the Gardners thought that the young ape would pick up the language through imitation, an integral behavioural trait in wild chimpanzees. To accelerate the pro-gramme, researchers physically moved Washoe's hands to form the sign shapes. This was in no way forced but was instead carried out in the same way we teach very young children to eat with a spoon, by holding their hands and guiding them gently to the mouth. This is known as moulding and eventually proved to be the most effective and fastest way of showing Washoe the correct signs; pure imitation was of little use.

Painstaking records were kept of her progress and the first voluntary meaningful sign appeared three months after the project started. Within a year she was able to string together different signs to form cumbersome but intelligible phrases. The programme was abandoned after more than four years and by that time Washoe could use and understand 132 distinct signs.

Her usage of signs was both abstract and imaginative; the signal for dog would be used for a real animal, a photograph of a dog and the sound of a dog barking. This indicated that she knew what a dog looked like and had a fair concept of what a dog actually was. When faced with a swan, an image for which she had no signal, Washoe made the sign for 'water bird'. This was no accident, the chimpanzee had thought about the problem and arrived at a logical, articulate solution that could be expressed within the confines of her limited vocabulary. I saw the same creative use of language from my two-year-old son Alexander. On a visit to a doctor friend's house, he saw a skull for the first time. He stared in fascination for a while and finally took me over to see what he called the 'bone head'. Shortly afterwards he announced that he had found a 'ball map', which turned out to be a globe. Children and chimpanzees share many learning experiences that are not copied from their elders.

Interestingly Washoe made the same sort of mistakes as human children when they first learn to speak. Word order is a difficult concept to grasp and it takes a while before they follow the basic laws of sentence structure. Washoe never reached a level of full understanding. She would use the right words but string them together in a way that made little conversational sense. 'You me out' was a request to go outside: this later evolved into 'You out me.' The meaning was present but it was phrased in a strange way that never really improved despite intensive tuition.

Notes for scientific use were made only when the chimp was in the company of humans, but she was often watched while completely alone at night. The Gardners were surprised to see that Washoe signalled to herself, in a way that made sense. When scurrying off to the potty she would use the sign for 'hurry'. While looking at pictures in a magazine she would form the signs for familiar images as they cropped up, even though there was no one around to interpret. She was talking to herself.

A subsequent Gardner project attempted to take this programme further by starting with a younger animal. We know, through studies on humans, that the development of language may be stunted if a child is deprived of hearing speech from birth. At this age they certainly have no idea of conversational content, but they listen and absorb intonation and speech patterns. By the time they reach four months old, babies are believed to be able to differentiate the sounds and rhythms of their native tongue from those of other languages. This prompted the Gardners to start again with four new-born chimps.

Progress with this group was much more rapid than it had been with Washoe and, as other chimps were present, there were expanded opportunities for learning and application of language within a peer group. These young chimps, and others in later study groups, were seen signalling between themselves about matters dear to their hearts such as food and play. Most of this interplay was strictly between chimps and was totally independent of human participation or interference. Much of the sign language was grammatically correct and used in exactly the same context as a chimp would have signalled to a human. On one occasion a chimp asked another for a banana by name. This was probably the first time in history that an animal had made a specific request to another animal using a structured language. Chimps will beg from each other in the wild but only in general terms and the actions used are similar to those requesting reassurance. The implications were stunning.

Instinctive behaviour is very powerful and difficult to override. Many researchers were interested to see if the apes themselves would opt to use a learned language out of choice. A further experiment was set up to look at the possibility of adults passing on signing skills to their offspring. Washoe was allowed to mate with an ASL-trained male called Ally and she soon became pregnant. Female chimpanzees develop very round abdomens during the final stage of pregnancy and this prompted Ally to signal 'What in your stomach?' to which she replied 'Baby'. This was the stuff of dreams for those researchers who witnessed the interchange.

Sadly Washoe's infant died at two months old. Immediately after the death Washoe signalled to her human handler 'Baby?' She received the reply 'Baby dead. Baby finished.' Washoe showed all the signs of depression, a state experienced by female chimps in the wild under similar circumstances. To compensate for the loss a ten-month-old male called Loulis was brought in and presented to her. Washoe was extremely excited when she was told beforehand that a baby was coming, but she was less impressed when it turned out to be a stranger rather than her own off-spring. Loulis was equally unmoved. He was very wary of Washoe and evaded her overtures of play and attempts to pick him up. Later, in the dark hours of early morning, Washoe deafened the youngster by beating her arms noisily against the metal sides of their house. Loulis was badly frightened by this and sought refuge by rushing into Washoe's arms. This broke the ice and the two quickly settled down together.

Loulis imitated his first sign eight days later. By the age of two and a half he had acquired a vocabulary of seventeen signs, only one of which appears to have been learned from a human. The others were copied from his mother and another ASL-trained female. Washoe even moulded Loulis's hand signals for important signs such as food, just as she had been taught. At the age of six Loulis had mastered fifty-five distinct signs.

The potential applications of human–ape communications were thought to be boundless, particularly in the field of animal behaviour. One hypothesis explored the possibility of taking a language-trained chimp back to its native habitat. Could the ape then act as interpreter, passing human thoughts and questions to the wild chimps and vice versa? This might be the doorway to true inter-species understanding and would be infinitely more profitable than years of conventional research.

Training an ape to the point where it can express itself at any depth requires a massive investment of time and effort from a very early age. The young animal spends so much time with its teachers that it effectively becomes humanized. Attempts have been made to introduce ASL-signalling chimps to zoo animals; the results were very mixed. Chimps that had had previous contact with their own species quickly reverted to their natural communication of facial expressions and vocal sounds. Those that had been with humans from a very early age either ignored the other chimps or ran off because they were frightened. A small number treated them as humans and attempted to communicate with them in sign language. Unfortunately, the chimps that were best equipped, through their intensive training, to feed back information to researchers were the very same animals that could not relate or communicate with their own species. One trained male, Arthur, was truly mystified over the nature of the chimps he visited. He simply called them 'black things' and took little notice beyond that. The use of trained apes as a bridge between species has so far been completely unsuccessful.

Because language training is so time-consuming, there have been very few programmes on which to base any concrete conclusions. There is obviously an element of luck involved in the selection of the individual animal. No one knows for sure whether Washoe was, linguistically speaking, an intellectual giant or possessed learning difficulties. There are just too few other cases with which she can be compared. The success of any project rests not just on the apes' brain power but also on the skills and resources of the teachers and the methods they have chosen to use. All these factors help to thoroughly muddy the water.

Since the days of Washoe, other programmes have come and gone with varying results, but there was an interesting development when the ASL approach was replaced by a form of written, or to be more exact symbolic, language. In 1972, at the Yerkes Regional Primate Centre in Georgia, USA, a two-year-old female chimpanzee called Lana was taught to use a computer keyboard that was not fitted with conventional letters but geometrical shapes known as lexigrams. The keyboard was wired directly to a computer programmed to recognize commands and 'sentences' formed by a string of up to seven lexigrams.

Lana was shown what the symbols meant and how to put them together in an order that made grammatical sense. The chimp could communicate with the computer, and its peripheral operations such as a projector system that supplied visual images, in order to ask for a drink or an apple. If the 'word' order was acceptable to the computer, Lana would receive her food. The idea behind the use of a computer was, first, to eliminate the possibility of a fallible human misinterpreting the evidence and secondly, to prevent Lana reading or broadcasting body-language. It also had the advantage of being able to accurately store every interaction for later analysis.

Within the narrow confines of the symbol system, Lana learned to communicate and even ask questions. She showed a reasonable grasp of grammar and could hold limited two-way conversations with the researchers through the keyboard. But the remote nature of the system seems to have defeated an important objective: Lana understood the mechanism well enough but tended primarily to use it for self-gratification. The main aim of her conversation was to ask for – and receive – treats. The inherent impersonality of the technology seems to have repressed the curiosity and feeling for social communication that was apparent in the ASL programmes conducted with real people.

Language experiments have not been exclusively confined to chimpanzees, although they have been the favoured subject, not just because of their more manageable size but also because they exhibit so much gestural communication in the wild. To minimize group violence among the highly volatile chimps, they have evolved a complex set of body-language signals which greatly outnumber those used by the equally sociable gorillas (they are virtually unknown among the solitary orang-utans). It was thought that as they had a natural inclination to language, chimps would respond to training better than their cousins. Subsequent researchers have shown that, if anything, bonobos have an even greater capacity for mastering language. Some go as far as suggesting that the stream of squeaks and other vocal sounds used by bonobos may be a precursor of true language.

Later experiments, based on ASL, show that gorillas can be brought to the same level of two-way understanding in around the same time as chimps. Their grasp of language appears to be very similar. In 1992 a one-year-old lowland gorilla known as Koko became the pupil in a teaching programme run by Francine Patterson. For almost a year Koko was trained at San Francisco Zoo, before being moved to a more suitable and private pen. In a large wooded enclosure Koko was taught ASL, but this time Patterson and the other researchers spoke relevant words at the same time as the signs were made. This technique was called simultaneous communication. By the age of four Koko had used over 200 different signs and showed an undeniable skill for understanding the spoken word, even when it was not accompanied by ASL signs.

Just like Washoe, Koko turned to ad-lib descriptions for those images that were outside her experience. She referred to a toy zebra as a 'white tiger', while a cigarette lighter was a 'bottle match'. Both were inventive uses of her limited vocabulary. Patterson records that Koko later used language to lie, insult and even make jokes. Michael, a male gorilla that joined Koko in the experiment, is reported to have used signing to tell the story of his capture from the wild. Over a series of conventional tests, Koko produced an average IQ reading of 80. IQ testing is a controversial enough subject in the assessment of human intelligence; any findings, high or low, should be treated very warily when applied to an ape.

Chantek, a male orang-utan, was also taught ASL and mastered the use of 150 gestural signs. This came as a slight surprise, as it was always assumed that orang-utans probably have less of a genetic predisposition for language due to their solitary behaviour and the subsequent lack of need for structured communication. Yet they seem just as able as the other great apes to learn the basics of grammar under similar test conditions.

There have been several serious attempts to evaluate the artistic qualities, if any, of an ape's mind. Paul Schiller at the Yerkes Laboratories did some early work on the drawing ability of apes and this caught the imagination of Desmond Morris, who was then Curator of Mammals at London Zoo. He introduced the idea of drawing to a young male chimp called Congo. The ape needed little encouragement and was hooked from the very beginning; he would fly into a screaming tantrum when his drawing equipment was removed for any reason. After a few initial chaotic scribbles, Congo developed his own style of pencil drawing, much of it based upon a pattern of diverging lines arranged in a fan shape. There were many variations on this typical chimpanzee theme, but Congo also produced the series of dots, circles and loops that many young children create when they first put pen to paper. Over two years Congo's output was a prodigious 384 drawings and paintings.

The chimp was excited about all practical aspects of art and was particularly keen when allowed to use coloured paint. Other drawing apes had simply used their fingers to daub paint on to paper, but Morris felt that this was too uncontrolled and random. Congo was encouraged to use a brush and quickly learned to paint with it just as a human would. He preferred to paint with one colour at a time, which he applied in blobs or lines. If the colours were swapped, he added the new one to fit in with the existing pattern instead of just painting over it. Eventually there came a point when Congo felt the work was complete and would not add anything else.

The idea of ape art was born and in September 1957, together with work from Betsy, a chimpanzee at Baltimore Zoo in the USA, a two-chimp exhibition was mounted at the Gallery of the Institute of Contemporary Arts in London's Mayfair. The event was greeted with ridicule and derision before it opened, but many viewers changed their opinions on seeing the exhibits. Some newspapers reviewed the private showing exactly as they would an exhibition by a human artist. The *Daily Telegraph* wrote: 'The originality and freshness of Congo and Betsy are unquestionable.' Others were less charitable and poured scorn on the whole event, claiming it was an insult to both art and animals. One critic in the *New Statesman* concluded that the hostility towards ape art was just as biased as that felt by Darwin when the nineteenth-century establishment was scandalized by the suggestion that men and apes were actually related.

There was a huge amount of public and professional interest in the exhibition and the gallery was inundated with offers to buy the paintings. All were refused as no one wished this experiment to become an excuse for commercial exploitation. Congo's art attracted extreme views: the President of the Royal Academy was apoplectic in his public attacks on the whole exercise, while Pablo Picasso had unbounding enthusiasm for the work and went as far as hanging an original Congo painting on the wall of his own house.

Zoologists and philosophers have long discussed the nature of the intangible factor that makes humans different from apes. Physically we are very similar, so the distinction must lie elsewhere. Many people have suggested that it is our grasp of abstract thought that raises us above the level of apes, and this is particularly obvious in the use of language and cultural fields such as art and music. Early man certainly left concrete evidence of his interest in art. Life must have been hard and hazardous for members of these Stone Age cultures, but they still found time to express themselves in visual forms. Cave paintings from sites around the world indicate that art was an early intellectual pastime. But as apes are experimenting with these hallowed subjects, the dividing line is becoming less defined – though it can be argued that as wild apes have never been known to show artistic tendencies, it is exclusively human influence that has promoted the activity in captive animals. Apes show no ability or inclination for representational painting, but is that important? If it is art that supposedly separates us and apes, the lines may have to be redrawn.

Complex tests have been designed to discover if apes have any self-image. Temporary dyes were used to mark various parts of chimp bodies and the animals were then given access to mirrors. To cut a long story short, the chimps quickly realized that the mirror ape was an external view of their own bodies. If the mirror chimp had a strange new dye mark on its head, the chimp would touch the corresponding site on its own body. In other words, they recognized themselves in a mirror. This shows a remarkably high level of self-awareness. Very little detailed work has been carried out along these lines with other species, but most behaviouralists feel that an abstract idea of self-appearance is extremely rare. Conjecture without evidence is always a risky business in the field of animal behaviour, because so many ideas have been disproved in the past. However it is likely that most animals have no concept about the fact that they are a three-dimensional being with an external appearance. Chimpanzees have proved beyond doubt that they have a sophisticated self-image. Orangutans show a similar level of mirror recognition, but the jury is still out on the gorillas' ability to identify themselves.

TOOL USE

The application of intelligence is so varied that it can appear in almost any form. Some individual apes show a definite skill with language while others solve complicated technical puzzles. There is one area in which the apes excel above all other animals, with the single exception of humans, and that is the use of tools. Only a handful of species are known to use inanimate objects to achieve their aims. Egyptian vultures use rocks to break open eggs and sea otters hold flat stones to their chest against which to break open shells. Although these are tools at their most simple, their use represents a major step forward in feeding behaviour. Both otters and vultures are mere beginners with tools when compared to the true masters, the chimpanzee.

Gorillas have been seen, and taught, to use tools in captivity, but by general consensus it is thought that their rich natural habitat allows them to gather food so easily that there is probably little need for them to resort to artificial feeding aids. Similarly, orang-utans can pick up basic tool manipulation with the help of intensive human tuition.

There is one fascinating aspect of the behaviour of wild chimpanzees that has only recently been thoroughly investigated and that is the importance of group culture. The ASL-trained chimps could communicate because they had been taught and exactly the same happens in wild populations. Some groups have learned tricks that are unknown to a group living 160km (100 miles) away, and even within a group individuals can teach themselves ingenious tricks that are unique. At Gombe Stream in Tanzania, chimpanzees were once baited with bananas as a way of bringing them to the study site. Bananas are a favourite food of chimpanzees and they had to be kept in strong boxes to prevent them being stolen.

One young male worked out how to open the box by unscrewing a door handle. At first, when he reached the prize, he hooted with pleasure and excitement. This attracted the bigger males, who muscled in and ate the lot. But they could not master the tricky business of actually getting to the bananas. Later the young male was seen breaking and entering, but he timed his raid carefully and did not even attempt to seize his booty until he was completely alone. The extent of his learning did not become apparent until the researchers saw him look around to make sure that no other chimps were watching before taking an armful of bananas. This time he did not make a sound. Chimps always express excitement noisily but this one had soon learned that if he called out, his ill-gotten gains would be stolen by bigger males.

Although chimpanzees have an impressive capacity for inventive behaviour, they are also great mimics and will copy actions that are profitable or interesting. A high-ranking chimp in a zoo once received an injury to a foot and was forced to limp noticeably for several weeks. Lower-status animals were fascinated by this unusual way of walking and soon copied the action; they even favoured the same limb. This strange behaviour spread and soon a fair percentage of the group appeared to have nasty leg injuries. This copy-cat behaviour continued well beyond the time when the original wound had healed and the patient had lost his real limp.

A talent for mimicry is a vitally important asset when chimps first start to use tools. Babies learn by watching their mothers and first attempts are usually pitiful. One of the simplest of all chimpanzee tools is the anvil which, once mastered, is a useful way of breaking into hard-shelled nuts. The nut is rested on a hard surface and hit sharply with a hammer made of stone or wood. A skilled adult can crack a shell with just a few blows, youngsters have more difficulty and find an almost endless variety of ways to get it wrong. They rest the nut on a rock and energetically use a branch to thump the ground beside it, or they put the nut on a soft surface and succeed only in driving it into the ground.

Hammer stones can weigh as much as 20kg (44lb) and, as stones are rare commodities in most tropical forests, a good one is stored to be used

over and over again. Chimps have even been known to carry hammer stones to good nutting trees, rather than keep ferrying nuts backwards and forwards to the anvil site. A good nut opener does not rely on brute force; there are subtle procedures to be learned. The shell needs to be hit with enough power to crack it without demolishing the contents. Nuts are often hit at an angle, which deflects the blow and avoids smashing everything. The first strikes are powerful, until the shell starts to splinter, and then the rock is used more delicately to chip away until the centre is exposed.

Obviously chimpanzees can utilize only objects that are found inside their territory, which is why sticks are so important as tools. Big branches are sometimes leaned against trees and used as ladders; they are deployed as hooks to reach objects that are out of reach. One ingenious captive chimp even used pieces of wood as a series of pitons, which he carefully stuck into a wall and eventually climbed. Another used thin sticks as dental floss to clean between her teeth, an activity that has also been seen in wild populations. Sticks make excellent levers for prying open well-protected food like monkey skulls and they are first-class weapons for attack and self-defence. Adult males have the role of defenders/attackers and because they are impressively powerful; chimpanzees armed with thick branches are a threat to be taken seriously by any predator, including man.

Termite mounds pepper the landscape of many chimp habitats and each contains a huge colony of insects that offers a rich and tasty supply of protein. Some mounds are soft and the chimps simply dig out their meal with their hands. Other termite mounds are concrete-hard and must be breached with more subtlety. Each mound has dozens of small entrance holes which the tiny nocturnal occupants seal up during the day. To reach the termites, a chimp must find an opening and insert a thin stick.

The choice of stick is critical, it must be green and supple so that it can be pushed through the tunnels without breaking, which means it must also have exactly the right diameter. The stick is pushed slowly into the hole and then wiggled. This infuriates the termites, which defend themselves by biting. With a swift and steady pull, the chimp pulls out the stick and, still clinging to it, the attacking termites. These are quickly eaten and the stick is then pushed back into the hole. A hungry chimpanzee will often spend two hours working at this food supply. Termite fishing is a technique that requires practice. Adults will sit and wiggle sticks for the best part of an afternoon, while youngsters often try for a short time before they get bored and wander off. The difference in the attention spans of adults and children will be familiar to anyone with young children. My own are like butterflies that flit from toy to toy, spending just a few minutes with each. Concentration is a virtue that seems to appear with age in both humans

Chimpanzees use leaves to hold the stems when feeding from thorny plants.

and apes. But there is also a difference between the genders of chimpanzees. Females spend much more time termite fishing than males, but are less likely to actively hunt and kill large animals for food.

New ground was broken at Gombe Stream when Jane Goodall watched two adults, Goliath and David Greybeard, collect fresh green twigs for a termite-fishing session and prepare them by stripping off the side leaves that would get in the way. After several attempts at probing the hard termite mound, the end of a fishing stick becomes frayed, so the chimp tidies it by biting off the tip. It might seem trivial to a human, but this example of animals actually making tools was the first ever known to science. It was a major discovery.

Tool-making has now been witnessed in several forms. One of the most ingenious is used to collect water that is out of reach of a chimp's mouth. The animal simply chews up a large leaf until it becomes a soggy mess and then drops it into the water, where it acts just like a sponge – a leaf that has been crushed absorbs far more water than one with its tough, waterproof outer membrane still intact. The chimp lifts the leaf, sucks out the water and carries on until its thirst is quenched. Leaves are also used as toilet paper, to wipe blood from wounds and to scrape up sticky food such as honey. There is no doubt that chimps are highly inventive in their use of tools.

Observation of wild bonobos suggests that they do not use tools anywhere near as often as common chimpanzees. Because their diet contains a larger percentage of fruit, there is little need to use mechanical aids to find their food. However, a detailed survey in Antwerp Zoo, Belgium, listed a catalogue of sixty-two different actions that involved the manipulation of objects. Not all can truly be described as tools and some of the actions were obviously based on curiosity and play, but a few were just as creative as those of chimpanzees. Bonobos have been seen using leaves as sponges, wielding clubs and digging with sticks. At the same zoo on a scorchingly hot day I watched a female collect several leafy branches to hold over her head while she crossed the enclosure to reach some food. Being animals of the dense forest, bonobos are susceptible to sunburn and I can only guess that the female was using the branches as a parasol to keep the sun off her head and back.

While chimps are the unchallenged masters of tool use, they are not alone in their skills. Orang-utans can use objects creatively, but because so much of their time is spent in the trees, where all four limbs are needed for climbing, their hands are usually less available for manipulative purposes. Chimps spend more time on the ground, where their hands are free for exploration and invention. This is underlined by the fact that orang-utans in captivity, where there is less opportunity for climbing, show a real talent for tool use. Semi-wild orang-utans in rehabilitation centres (see Chapter 11) show more interest in tools than their truly wild counterparts, but there are two possible explanations for this difference. First, humanized

Orang-utans in rehabilitation camps readily use tools. These are probing the ground with twigs in search of grubs.

orang-utans spend a greater percentage of their time on the forest floor, probably because their human role models are earth-bound. Secondly, their close proximity to humans gives them the opportunity to copy our own use of tools. Orang-utans are great mimics and will quickly learn to copy any action that is to their benefit, whether it is performed by a human or an ape.

Rehabilitated orang-utans use sticks as insect probes and levers and many include human artefacts in their mechanical exploration. I once watched a young male in Tanjung Puting National Park, in Indonesian Borneo, carefully move a small canoe from its mooring, swing it 90 degrees around and use it as a bridge to cross a small but deep stream. The manoeuvre was polished and swift, leaving me with the distinct impression that this was not the first time the animal had taken this short cut. Wild orang-utans sometimes break off leafy branches to use as parasols to keep off the bright sun while resting high in the canopy – even orang-utans can get sunburned. And when feeding in trees protected with thorns, they wrap thick leaves around their hands to reduce possible injuries as they climb to reach the fruit.

As all authorities admit, the more we learn about the apes' biology, the more we must constantly reassess our opinions of their intellectual capabilities. Almost every new discovery suggests that the great apes are more adaptable and inventive than was once thought.

8

THE COMMERCIAL APE

OVER THE PAST thirty years travel has become remarkably easy and relatively inexpensive. Tourists regularly cross the world to enjoy a few days' sunshine in the gloomy depths of winter. As travel has both commercial and status-confirming implications, there is a constant battle by all participants to push forward the boundaries and search for new experiences. The recent 'green' boom has created a huge number of tourists who wish to see for themselves those exotic and dramatic animals that they watch on television every day.

A lucrative new market has grown up around wildlife-watching as a leisure activity. Inevitably, the large, charismatic species are the holy grails for the majority of eco-tourists – as they are known in the trade. Advertisements for trips specifically to see great whales, tigers and polar bears can be seen in countless glossy magazines. For many people, the ultimate wildlife experience would be to see the mountain gorilla. Following appearances in beautifully made wildlife programmes on TV and expertly marketed feature films, these apes have taken on the aura of a modern myth.

There are three countries where it is possible to see mountain gorillas: Uganda, Zaire and Rwanda. The first two offer excellent opportunities, but for many true enthusiasts Rwanda is the only possible destination, for this is where Dian Fossey carried out much of the work that brought gorillas to the world's attention, and, to be more practical, the gorillas here are the most approachable. There is an infrastructure in place that permits visitors to see the apes, but it is cumbersome and expensive. Aspiring gorilla-watchers need to be determined, fit and well heeled. The starting point is the small town of Ruhengeri in north-east Rwanda. From here a four-wheel drive vehicle must be taken to the headquarters of the Parc National des Volcans, where visitors are allocated their place.

Mountain gorillas in Rwanda are an excellent example of how endangered animals can generate money to aid their own survival programmes.

Each group can be visited only by six people per day, although it is rare for all places to be taken. The cost of looking at the gorillas for just one hour is around $120 per person. Once the fees are paid, a vehicle is taken to the park boundary and the rest of the journey is on foot. At 4,000m (13,000ft) above sea level, the air is thin and many people turn back before seeing even a gorilla footprint, due to breathing difficulties or even altitude sickness. There is one slight advantage to wildlife-watching this far above sea level and that is the lack of mosquitoes. The experience of walking in low-lying rain forests at certain times can be made unbearable for tourists by the unremitting bites of these insects, but this particular discomfort is rare in the Virungas. Cloud forests are surprisingly quiet places, with little in the way of visible wildlife. On my first visit I saw no birds at all; the only moving life forms were small clouds of tiny flies. The guides lead each group to the site where the gorillas were last seen and then simply follow the trail through the forest. These are full-time trackers and they are very good at their job. The daily movements of each group are monitored even when there are no tourists. Experienced guides can find their apes under any conditions. There are several groups of gorillas that have been deliberately habituated to tourists, but they move over large areas and are sometimes out of range. If the visitors are fortunate, gorillas will be feeding in the bamboo forest on the lower slopes of the mountains. Walking is easy here as little grows between the giant bamboo stalks and the land is comparatively flat. On a good day the air is cool and the trek is pleasant.

On the other hand, the gorillas may have moved to higher ground up among the huge hagenia trees. This is true tropical woodland and the vegetation is often so dense that it is impossible to see more than a metre in any direction. Thick gloves and trousers have to be worn to fend off the stings of the massive nettles that blanket much of the forest floor. It is also important to be prepared for changes in the weather. In cloud forests conditions can alter dramatically in minutes. Torrential down-pours are perfectly normal in the Parc National des Volcans; conditions often get so bad that visibility drops to zero. Everything becomes sodden almost immediately and the vegetation may be wet for hours afterwards. For many, even if the rain doesn't dampen their view of the gorillas, they might encounter low, dense cloud that shrouds the whole mountain. The gorillas can always be heard as they feed, but frequently they are seen merely as grey uncertain silhouettes through the mist.

The first sight of the group will be a fleeting glimpse – black shapes in the undergrowth ahead. Then, as the gorillas pause for one of their feeding stops, the tourists eventually catch up. At first the interest is mutual: the gorillas immediately recognize the guides and researchers that they see each day, and are fully aware that the other humans are new. The gorillas sit, watch and feed at the same time. Youngsters are quick to pluck up courage and often run by on all fours to brush visitors with the back of a

The accessibility of humanized mountain gorillas has turned the species into media superstars over the past twenty years.

hand before hurrying off. Unguarded camera bags and coats will be grabbed, emptied and thoroughly explored before either being thrown to one side or ripped to pieces. It has been known for some time that gorillas respond to people in different ways: one day they will show fear of a group while the next another set of strangers will be examined, poked, prodded and even followed. There seem to be no rules to this change in attitude, or, to be more precise, none that we understand. The apes themselves obviously detect some aspect of the visitors that they find either interesting or threatening.

Dian Fossey's early work shows exactly how to approach gorillas without alarming them. Before closing in, visitors are given a strict set of behavioural rules by their guide. No eating or drinking, as both activities may arouse too much curiosity. No eye contact, particularly with adult males, as this is the first sign of outright aggression with gorillas. Not only should eyes be turned away but, ideally, the whole head should look down. If a male charges, drop to the ground and curl into a tight ball. And, finally, no one – no one at all – should approach or touch a gorilla. Physical contact regularly takes place, but it should always be the gorillas that instigate the action.

The dominant silverback is always close at hand when tourists visit but usually keeps to the edge of the group and feeds less than the others. Watching for potential threats or challenges, he sits quietly, constantly flicking sideways glances at the strangers. It is then that first time visitors realize exactly how huge male mountain gorillas can be. They are quite simply vast; their sloping backs could be used as a fair-sized table and their forearms are thicker than most men's thighs. Another lingering first-time impression is the distinctive warm, musty, almost farmyard smell given off by gorillas. It is certainly not an unpleasant odour, but it is decidedly unique to these apes and not easily forgotten.

Sometimes visitors break the laws of gorilla etiquette and the silverback is pushed into a display of dominance. Display behaviour is strictly for show, used to maintain the family bond and reinforce individual status within the group. An irritated senior male generally treats a human in exactly the same way as he would a rival or subordinate ape that needed to be kept in line. The display may start in a low-key manner. In fact, the presence of a large male may be totally unsuspected by visitors until a nearby tree begins to shake violently. Next come the rapid, hollow slaps of chest beating and perhaps a short sideways rush.

This is generally the end of the performance and the male then goes back to watchful feeding, but charges are not unknown. Should this happen, it is essential to ignore the initial overwhelming urge to run – which would be pointless anyway, as a gorilla could quickly outpace a human. Behaviour of wild animals needs to be analysed from their viewpoint, not ours, if it is to be understood. The silverback charges as a way of intimidating potential enemies, and faced with this unavoidable challenge the antagonist can either fight back or submit. The acknowledgement of a higher-status animal is signalled by gorillas through a submissive gesture, with a lower body position and the avoidance of direct eye contact. These actions from either a gorilla or a human should pacify the silverback and calm the situation. Remembering the rules and sticking to them are real tests of character when a novice has to hold ground in the face of a charging gorilla. Occasionally, an adolescent will run by and slap a visitor's leg with the back of his hand. This can be an uncomfortable encounter but leaves nothing worse than a pink handprint on the skin.

No human in his right mind would attempt to fight off a charging gorilla. The only other alternative to submission is running, which could aggravate the situation, because it breaks the unofficial gorilla code and is outside the normal rules that govern behaviour within the group. This can result in a real attack. Crouching low, with a bowed head, indicates to the male that the rival accepts his superior position. Unrepentant retreat, however, means that the adversary does not want to be hurt but may still pose a challenge to authority. This must be handled swiftly and finally if the silverback is to hold rank. Rival males or careless humans are treated in the same way. Gorillas use their hands and teeth in fighting, and both are capable of killing a human. All reliable sources agree that reported attacks by gorillas have always been the result of self-defence or foolish behaviour on the part of humans rather than pure aggression from the apes. Dian Fossey tells of a young tourist who attempted to pick up a baby gorilla, even though the whole family group screamed out warnings. Before he touched

Wardens in the Virunga national park keep a close eye on the resident gorilla groups, but their low numbers make it impossible to police the whole area.

the youngster, the mother and silverback attacked. The unwise visitor survived but still bears the scars.

Tourists rarely see this aspect of gorilla behaviour. The majority spend their allotted time with the family group and go home carrying memories of a magical hour that will stay with them for life. Some are genuinely lucky. On my first day in the park, I was constantly followed by an adolescent female who kept touching the back of my legs whenever I was occupied with something else. While lying on my stomach flat on the forest floor, keeping low down to photograph the silverback, she climbed on to my legs and sat for several minutes. When I turned to look, she slid off, rolled over and over, and finally disappeared into the bush. Close contacts like this are far from unusual. The gorillas are curious about visitors, particularly as they tend to meet towards the middle of the day, when the pace of feeding has slowed down and the youngsters are bored and looking for distractions. Touching is usually brief, but some tourists are given the full treatment and find themselves acting as a climbing frame for a handful of young gorillas. They might also experience mock charges when a silverback runs directly at them and veers away at the last moment.

The use of gorillas as a tourist resource has led to much debate among conservationists; there are two diametrically opposed schools of thought. The opening up of such an endangered species to public view poses real threats to the gorillas. Being closely related to humans, they are susceptible to many of our diseases, including pneumonia and colds. Unfortunately, they lack the genetic resistance that has been built up by man and a simple dose of flu could be fatal to a gorilla that had no antibodies to fight it. The Parc National rules state that no one showing signs of illness may visit the gorillas, but this seems to be flexible – to say the least – and even if the rules were observed, some diseases are at their most contagious before any symptoms are visible. A visitor could introduce a virus without being aware that he carried it. An alien strain of pneumonia would sweep through a family group in weeks, leaving a trail of dead animals in its wake. Remote groups of gorillas show a natural wariness of man and require many months of painfully slow assurance before they accept our presence. Habituated groups are almost blasé in front of tourists; the loss of this instinctive fear leaves them highly vulnerable should they meet humans who wish to take more than photographs. It has been argued that the 'approachability' of these gorillas has made them easy targets for poachers. But a high-profile existence could also be seen as a built-in safeguard; the very fact that they are so public and could be visited at any time by a group of tourists with a ranger may deter all but the most determined poachers. The gorilla wardens in Rwanda are a dedicated bunch who appear to believe deeply in their work. Poachers risk death if caught in a shoot-out with Parc National guards. Unfortunately, the guards are often less well armed than poachers, as the profit from selling black-market gorilla goods is much higher than the budget for running the park.

To be absolutely fair, Rwanda, Uganda and Zaire are relatively poor countries that cannot afford to meet all the needs of their human population, let alone those of their wildlife. In the West, where we already have good roads, hospitals and schools, conservation suffers through massive lack of finances. Much of central Africa is deprived of even basic commodities such as clean water and primary education, so, economically at least, conservation is not a major issue. These countries just do not have the resources to spend on anti-poacher patrols. But if the gorillas can be used to generate income, particularly in the form of hard foreign currency, they then become an asset to be protected by both the government and the locals, who might benefit from the creation of new tourist-related jobs.

In Rwanda the mountain gorillas are the nucleus of a growing business. Visitors are not only charged to enter the Parc National des Volcans to see the gorillas; they have to part with money to stay at hotels, eat, hire drivers and guides, and buy souvenirs. Although the tourist trade has suffered terribly as a result of Rwanda's political instability and on-off civil war, the gorillas are now a major industry that has enormous potential if carefully nurtured. The country has a vested interest in taking care of the apes, if only to protect its income. With all due respect, Rwanda has little else that would tempt a tourist so far from the beaten track; if the gorillas disappear, so does the cash. This may not be an ideal form of conservation, but at least it is practical. The country needs the gorillas and, without help and sympathetic management of the park, the gorillas would die out. If these creatures lived in the West, there would probably be an entire industry built around them. Luxury coaches would ship people back and forth on an hourly basis, stopping off on the way for the *Son et Lumière* show and a chance to buy inflatable gorillas. The danger is that eco-tourism changes from a pro-conservation project into a purely commercial venture that ends up destroying its only point of interest.

Similar schemes are now appearing in South-East Asia, with orang-utans as the star attraction. In both the Indonesian and Malaysian sectors of Borneo, tourists can pay to see orang-utans. The approach to eco-tourism here is less structured and commercialized; it is also far less expensive than gorilla-watching. But the philosophy is similar: the apes are earning money for their own conservation and the host country.

A primary tropical rain forest is dark and extremely dense, and most first-time visitors find its humid and sticky heat very uncomfortable; some never become accustomed to it. Fuelled by countless TV films, rain forests have acquired the reputation of being chock-a-block with animals. This is true to a degree, but most tourists see nothing more than the armies of ants that constantly search the forest floor. Mosquitoes and leeches are relentless in their quest for tourists' blood and, in this perpetually damp habitat, both are found in huge numbers. To call them irritating is a ridiculous understatement; these creatures can be so persistent that many people leave the forest vowing never again to return.

Rehabilitated orang-utans are more terrestrial and approachable than their wild counterparts.

There is wildlife other than invertebrates around, but most are hidden high in the canopy, are completely nocturnal or are frightened of humans and avoid all contact. Apart from the calls of strange and distant birds, rain forests are normally an empty, silent disappointment to the uninitiated. Being tree-dwellers, orang-utans are fiendishly difficult to find in the wild and, because they are solitary in their habits, a single orang-utan makes less of a show than a family of ten gorillas. Very few people ever attempt to find a truly wild orang-utan; average tourists can only realistically expect to see them in rehabilitation centres. The most famous, and accessible, of these is Sepilok, which is 25km (16 miles) from Sandakan, in north-east Sabah. Although there are several others, they tend to be in more remote forests that deter all but the most dedicated travellers. An aspiring ape-watcher would be inclined to think that orang-utans, with their sheer bulk and covering of shaggy colourful hair, are unmissable, but when feeding or sitting quietly, they blend into their surroundings to the point that they can be passed by without a second glance. Even when they move, orang-utans are more often heard than seen.

In rehabilitation camps there is no problem at all spotting orang-utans. They quite often drop out of the trees and on to the shoulders of eager visitors and nothing is safe from their strong, prying fingers. The tourists may go home slightly battered, with one or two of their possessions missing, but at least they have seen a 'wild' orang-utan at close quarters. Curiosity about humans keeps 'camp apes' on the forest floor more than wild orang-utans, and this makes them far more approachable. Some conservationists see this as exploitation, but there is no doubt that the money brought in by foreign visitors encourages governments to at least pay attention to the animals that are attracting a steady profitable stream of tourists.

Tourism may be contentious but it is often less harmful than other, more commercial, methods of raising money. In 1969 around one-third of Rwanda's Parc National des Volcans was officially commandeered and turned over to the cultivation of pyrethrum. From these plants a natural insecticide can be extracted that is much cheaper to produce than the usual synthetic varieties. The project was sponsored by the European Union (then called the Common Market) and was responsible for the destruction of prime gorilla habitat. Evicted animals were forced to move on to new areas, where they came into contact – and conflict – with resident groups. Pyrethrum is still an important crop in Rwanda, being one of the small number of plants that thrive on the high slopes of the Virunga Mountains. Every possible corner of land is turned over to its cultivation, even tiny patches measuring just a metre square. It is possible to stand with one foot inside the official gorilla sanctuary while the other nestles in a stand of pyrethrum.

Gorilla territory throughout Africa is coming under threat of destruction as the result of timber felling. Humans have cut down living trees

In Rwanda pyrethrum is cultivated to within centimetres of the national park boundary. And each year it pushes further forward.

The promise of food and stimulation encourages reintroduced orang-utans to stay close to camp. This animal is greeting a boat arriving at Camp Leakey in Tanjung Puting National Park, in Indonesian Borneo.

since we first picked up a sharpened stone, but while our population was low, deforestation took place at a sustainable rate. The demand for trees was negligible and a tropical forest could supply everything without being affected. Today deforestation is more like a military operation, with huge areas being flattened at any one time. Not long ago an area stripped of trees would have been left once the timber had been used and, given time, the forest would have naturally regenerated. In the twentieth century the human population is too high to waste such an opportunity. Now, once the trees have gone, farmers move in and shanty towns appear. The woodland has no chance of recovering.

The conservation of apes has to have a two-pronged approach. First, the forests in which they live must be protected from further destruction. Secondly the animals themselves need to be safeguarded against direct hunting. In both Indonesia and Africa there are local education programmes which aim to warn people of the dangers of deforestation and what will happen to their land as a result. There are also classes that introduce apes on a more personal level than perhaps they were viewed before. If animals are seen purely as an infinite resource, hunting continues without question. One way to challenge overhunting is to show potential poachers how their actions affect their prey, along with the forest and all of its inhabitants. Obviously, some people are completely impervious to education programmes such as this, but there is strong evidence that many become more protective of their wildlife when they discover that it is special. An enthusiastic and watchful local population of humans is one of the best defences that animals can have. And if the apes are the core of an industry that pays real wages, it is in everyone's interest to ensure that the animals prosper.

9
OF APES
AND MEN

BASED ON OUR knowledge of the appalling treatment humans have meted out to animals throughout history, it should come as no surprise that apes have fared little better at our hands than most other species. The only difference is that apes have been on the receiving end of the most complex and high-tech aspects of human behaviour. Once it was realized that the great apes were anatomically so close to us, many researchers saw an opportunity for experimentation in areas that were legally and morally impossible to explore in *Homo sapiens*.

Much of the early work was basic dissection just to confirm that apes were the ultimate research tool. This was quickly appreciated, but the use of laboratory apes proved not to be the hoped-for panacea, simply because of economics. Gorillas and orang-utans were never likely to be popular candidates for laboratory research because they are too big to keep and handle easily. Chimpanzees were more attractive for several reasons: they appeared to be the closest to humans anatomically, their smaller size required less space and food to maintain, and they were the most numerous species.

The apes may have ended up in high-tech labs, but the methods used to apprehend them in the first place were, and still are, crude. Foot snares and primitive firearms are used to catch or kill mothers with infants. Today it is possible to walk through many markets in Central Africa and see chained infant chimpanzees for sale. The mother, being totally unmanageable, would have probably been eaten.

Researchers had two ways of obtaining chimps: either they could buy animals from dealers or they could set up a colony and breed their own. Neither option was ideal. Most behavioural and physiological work on primates had, until then, been carried out with small species such as rhesus monkeys. These are comparatively simple to catch and transport; they also breed rapidly in captivity, with each female producing young every year. For only a small outlay, a steady flow of animals could be fed into the research system, but rhesus monkeys became old hat once chimpanzees were available.

Chimpanzees are immensely popular zoo attractions,. Fortunately, most are now kept socially in large enclosures that offer a degree of stimulation.

Chimps were up to 100 times more expensive to buy than small monkeys and their slow reproductive rate ensured that captive breeding colonies were costly to run for the small number of animals produced. Instead of apes taking over entirely as guinea-pigs, they became the final-stage subjects. Early work on drugs and surgical techniques was performed on the usual rats, mice and monkeys. When this was taken as far as possible, one or two last experiments were carried out on chimps to hone the procedure before it was applied to humans.

Some of the areas of experimentation that took place in the 1960s and 1970s are now seen as ethically suspect to say the least. Among many other questionable tests, chimpanzees were forced to undergo chain-smoking of high-nicotine, unfiltered cigarettes for hours at a time in an effort to understand the carcinogenic effects of smoke on human lungs and other organs. Chimpanzee brains were the closest available substitute for those of humans and they were seized upon for a huge variety of invasive investigations. Electrical and chemical implants, radical surgery and genetic manipulation all featured, among other procedures, in clinical ape research. Some of the larger pharmaceutical companies experimented on their own resident chimpanzee colonies and kept the findings secret. Even today we cannot judge the extent or content of those early projects.

Today the use of animals in any kind of harmful testing is highly sensitive. Relentless consumer pressure alone is often enough to steer big business away from live experimentation. A single chimp can now fetch up to £15,000 ($24,000) on the open market and the price is always rising. The soaring cost of apes will probably ensure that few end up in laboratories of the future. There are exceptions, though. Currently chimpanzees are the favourite subject for the penultimate tests on drugs to control AIDS. Being physically so close to us, they can be infected with the human HIV virus and then studied as they are given various new prophylactics and treatments. The company that eventually produces the definitive drug to control AIDS stands to make a considerable fortune; the price of test chimps, therefore, is a small investment compared to the possible profit involved.

A variation on medical experimentation was the use of chimps to predict the effects of unknown forces on the human body. Space flight and its potentially disruptive problems was an obvious field for chimp substitution. In 1961 Enos, a male chimpanzee, was the first live creature to orbit the Earth under the United States flag. He completed two successful orbits, while being fed with water and concentrated banana pellets. Enos was well strapped to his flight seat to stop him moving around and damaging the inside of the capsule. Part of the importance of this test was to gauge the effect of rapid acceleration and weightlessness on a near-human subject, so Enos had to be well looked after if the results were to be valid. To guarantee he ate during the journey, the chimp was fitted with electrodes that administered mild shocks unless he regularly pressed the lever of the food dispenser. This conditioned behaviour was thoroughly ingrained in the chimp before he left the ground.

Chimpanzees have long been utilized for our convenience and the rationale has not always been based around scientific research; all too often it is purely commercial. Visitors to popular beaches in Spain and a few other Mediterranean countries are regularly approached by local photographers trying to interest them in holiday portraits. Competition is fierce and, to attract more business, some photographers bring along an animal 'prop'. Holidaymakers can have their photograph taken with a macaw on their shoulder or a chameleon climbing up their arm. For some time the ultimate animal accessory was a baby chimp. As most people never have the opportunity to handle one of these undeniably cute creatures, they quite cheerfully pay to be photographed with a three-year-old chimpanzee, dressed in a sailor suit or baby clothes, nestled in their arms. Business can be brisk and profitable for photographers with such an attractive assistant.

Few, if any, of these chimps ever arrived on the beach by authorized routes, as their legitimate cost would have been prohibitively high. Most are smuggled in from Africa and bought on the black market. Smugglers have fewer scruples and less paperwork than licensed dealers, and this makes their prices lower. We know a lot about the methods adopted by poachers who live-trap wild chimps in West Africa, but we can only guess the number of animals that are killed in order to sell the one single infant that reaches the market.

Many die during transportation and, once they arrive at their final destination, life expectation is murderously low due to the ignorance of their new masters, the squalid conditions in which they are kept and the ill-treatment they experience at the hand of their owners. Young chimps are often drugged and packed into tiny boxes to keep them quiet and hidden during the trip from Africa. Sedatives are also used to keep them passive while they are handed from tourist to tourist. Daylight hours are long in the summer and humans get tired in the Mediterranean heat. When a photographer goes to rest or eat, the chimpanzee must continue working, for he is simply passed on to another photographer and hawked around the beach.

Young chimpanzees may not be as powerful as adults but, like most mammals of this age, they are very playful and energetic. In the wild, young chimps investigate the world with their mouths, just as human infants do. They chew out of curiosity and play. Customers of beach photographers are unlikely to enjoy the play bite of chimps, so before they ever reach the public many apes have their teeth removed. Judging by the appearance of confiscated chimps, their teeth are not surgically extracted but knocked out. They are frequently broken off at the gum, with the roots left in, and this can result in painful infections that would rarely be treated.

How many visitors, when handing over their holiday pesetas, realize that the chimp featuring in their family photograph album will spend that night, and every other of its terribly short life, locked in a tiny concrete basement 2m (6½ft) square or chained to the chassis of an abandoned truck? Chimps have been found in both of these prisons and there are even worse conditions that will never be discovered. The working life of these unfortunates is limited. Most die of neglect within a year or so, but as a beach photographer can recoup the chimp's purchase price within a month, that is a high profit margin on the original investment.

The few chimps that survive longer are commercially useless by the age of six, for by then they are too strong and self-willed, and so pose a real threat to both their handlers and the public. As they do not officially exist, unwanted chimps cannot be sold on to zoos or reputable dealers. Some must find their way back on to the black market, but there are few buyers for such large, unpredictable animals. We must assume that the majority are simply killed. They are then replaced by other youngsters that are doomed to an equally short and squalid life. The use of chimpanzees by beach photographers has now been banned in many areas and this has reduced the number of cases, but sadly the trade has not been wiped out. In an effort to clean up their eco-image, Spain and other tourist-reliant countries go to considerable lengths to track down offenders. Fines are heavy and prison sentences are not unknown. The chimpanzees are confiscated and passed on to qualified keepers or rescue centres.

The unpredictable nature of chimpanzees makes them unlikely candidates for commercial tourism.

Orang-utans are still kept as illegal pets in South-East Asia.

Animals that have been through this experience are usually physically and behaviourally damaged, leaving them no possibility of ever being returned to the wild. They must spend the rest of their lives in zoos. One figure estimates that as many as 6,000 chimpanzees are currently in captivity throughout the world, in zoos and circuses, or held illegally. As many as 1,000 more are taken from the wild each year to replace those that have died and to fill the demand from new clients.

In Asia illegally collected orang-utans are used not as beach models but as attractions in the night-clubs of Taiwan. Little imagination is needed to appreciate that their lives are no better or longer than those of Spanish beach chimps. Estimates vary from 600 to 1,000 orang-utans currently kept in Taiwanese night-clubs. There is a half-hearted campaign to stamp out the practice and a few animals are occasionally confiscated and passed on to rehabilitation centres, but the vast majority spend their miserably short lives in the oblivion of tiny cages and smoke-filled cellars.

This young orang-utan was shot while being captured. Her mother was killed and the baby developed gangrene as a result of the gunshot wound. Her right arm was amputated to prevent the infection spreading.

For obvious reasons, gorillas are not treated quite so carelessly as the other great apes; their prodigious strength even at a young age makes them extremely awkward to keep and handle. However, they are not exempt from ill-treatment, which just comes in a different form. In West Africa gorillas have the reputation of being crop-raiders, and this is true to a degree. Farmers with agricultural land bordering gorilla territory often don't wait for proof, shooting any gorilla that happens to wander anywhere near.

But hunting does not need an excuse. Every year up to 800 lowland gorillas are killed for their flesh. In the protein-hungry regions of Central Africa, almost any wild animal may be killed for food. This is known as bush-meat and is a valuable commodity for local people. A male or female adult gorilla is big enough to provide a small mountain of meat which will feed a family for some time. The population of lowland gorillas is probably sufficiently healthy to withstand a small amount of hunting by locals feeding their families, but recently the agenda has altered. Gorilla meat is now being offered on the menus of town restaurants and even more is eaten in the hidden depths of the forest. Commercial loggers have moved into gorilla habitats and they have a large workforce to feed. Bringing in food is slow and expensive; it is much easier to hire a local hunter to gather supplies from the surrounding forest. A gorilla can be killed with a single bullet and provides a lot of meat, so few hunters would ignore them as a quick and efficient way of earning money.

Snaring is an even cheaper way of killing, but it is less selective. Wire loops are left out in the hope of catching something big. They can trap or even kill many animals, but gorillas are strong enough to prise the wire away from its anchor. Unfortunately, they can rarely remove it from their limbs. Cutting deep into their flesh, the wire restricts the blood supply and becomes a site for infections such as gangrene. Ndume, a well-known male mountain gorilla, is a good example of the outcome. His right arm ends in a stump, showing where his hand was lost in a poacher's snare several years before. At least he survived, though. Many don't.

Food is not the only prize, as gorillas are also hunted for their reputed medicinal properties. The theory is simple. Many superstitious cultures believe that an animal's physical properties can be transferred to a human if the relevant part of the anatomy is eaten. This is known as fetishism and has been recorded for centuries all over the world. Owls' eyes are eaten in order to inherit their keen nocturnal eyesight and leopards' feet are said to endow a man with stealth and cunning in hunting. In Africa, where the gorilla represents invincible power, it is well known that if a man wishes to achieve a similar prowess, he must eat pieces of a big gorilla. Hands and arms are the most powerful fetish but, failing this, any part of the anatomy is believed to strengthen the user. In the Congo and several other countries, market stalls specialize in selling fetishes and openly display a range of gorilla bones, fingers and even heads, along with those of other animals. Business is healthy and apparently expanding.

Thanks to changes in the law and public opinion, few tourists would now attempt to take home trophies from endangered species, but only a few years ago it was possible to buy mounted gorilla heads as souvenirs in the most respectable shops. Probably the most sickening keepsake I've ever seen was made by cutting off a gorilla's hand just below the elbow joint. The dead fingers were permanently fixed in a knuckle-walking pose; the arm would stand on the floor, with a glass bowl set into the wrist bone. The finished item was sold as a free-standing ashtray, and even now these can be found in the homes of travellers from earlier decades.

With their near-human appearance and behaviour, apes have always been one of the biggest draws in zoos, and this popularity has been responsible for some of the most subtle and unintentional suffering inflicted by man. In the early days zoos were seen merely as places of public entertainment. Today the better ones go far beyond this, acting as conservation centres for education, research and the captive breeding of endangered species. Instead of keeping animals in concrete holes, curators of enlightened zoos provide large and stimulating environments for their animals. This is particularly important for apes, because their diet in the wild is often low in nutrition and they need to spend a large part of each day feeding. This involves climbing, travelling, jumping, stripping leaves and peeling fruit – tasks that keep them interested and

Jozi, a young mountain gorilla, was found in a poacher's snare. She was tranquillized by dart gun by researchers from Karisoke, but although the wire was removed, she later died of septicaemia.

active. In some zoos these same animals are given the equivalent of fast food, which is high in calories, delivered at the same time every day, easy to eat and gone in minutes. The apes then have nothing to do for the rest of the day.

The cost of keeping apes is such that some zoos can afford to keep only one or two animals. Combined with a catalogue of other depressing factors, social deprivation can affect the apes' mental stability. Robbed of both functional and social stimulus, solitary apes kept in stark cages invent their own ways of passing the time. Just for something to do they constantly regurgitate their last meal and eat it again, and again, and again. Faeces become a toy to be thrown around or smeared on the walls, and this atypical behaviour is regularly punctuated by a relentless pacing around the cage. The same pointless repetitive behaviour can be seen performed by humans who have been kept in similar understimulating conditions for too long. This mindless recurrent activity is known as stereotyped behaviour and is caused by sheer boredom. In extreme cases apes mutilate themselves by chewing their own fingers until they bleed. Good zoo directors now understand much more about this problem and new enclosures are designed to be as stimulating as possible. Responsible collections keep primates in pens equipped with climbing frames and provide toys and branches which are regularly changed to maintain the animals' interest. No *good* zoo would consider keeping solitary gorillas or chimpanzees, for they understand that these animals need complex social lives to stay mentally healthy. Twycross Zoo in Leicestershire, England, houses an excellent ape collection and its resident gorillas have their own colour televisions to watch, among other ingenious stimuli to keep them occupied.

Zoo economics have produced some unforeseen problems. Surplus primates are not always easy to rehouse. Many zoos have laudable ethics and won't sell their animals for experimental or commercial purposes. Yet there are few collections that can afford to buy and keep great apes. To combat this several long-running experiments are working on the use of reversible contraceptive techniques to prevent the overproduction of young apes in captivity. Subcutaneous implants and the oral contraceptive pill have both been tried with success. This is a programme that needs to be mastered for a variety of good reasons, but there is a terrible irony in the fact that we are trying to stop the breeding of species that are so threatened in the wild.

Gorillas breed successfully in zoos but their reproduction rate is painfully slow, just as it is in the wild. In the mid-1970s there were 497 gorillas known to be in captivity and of those 402 were wild-caught. At the moment, only lowland gorillas are held in zoos; mountain gorillas have never settled down to life in captivity, the majority dying after a very short time. Wild mountain gorillas have the limited protection of living within a national park, but the majority of lowland gorillas are found outside parks and other reserves. This may be of little importance, as in many countries the practical value of a legally protected area is open to debate. National-park status is not always as useful as it could be. It may not be too cynical to suggest that some national parks are designated in order to quell international protest rather than from a real desire to protect the environment.

Apes are highly vulnerable to the omnipresent effects of human activities and few non-natural catastrophes over the past decades have been quite as horrific as the near genocidal civil war in Rwanda. Following the shooting down of a jet carrying the president in April 1994, there was an explosion of violence between the rival Hutu and Tutsi factions. The two had long vied for political control of the country, but in the power vacuum following the president's death the arguments and skirmishing blossomed into a full-blown war and the killing reached a level that is almost impossible to comprehend.

One of the most politically sensitive areas of Rwanda was perilously close to the national-park home of the mountain gorillas, which lies on the border of Zaire. Many refugees and ex-government troops sought sanctuary from the slaughter in Rwanda by crossing into Zaire. The Virunga Mountains straddle the border and it was feared that gorillas would get caught in the cross-fire or be killed for food. At the time of writing, neither seems to have happened. But there were war casualties. The new government of Rwanda laid land-mines along the border in an attempt to prevent an attack from the regrouped forces of the deposed government. The mines are still in place, and at the time of writing pose a real threat to all local gorillas. The widespread chaos resulting from the war disrupted the fragile infrastructure that had been built up to care for the apes. At one stage the staff were forced to leave Karisoke, the research station that was at the heart of gorilla monitoring and conservation, as it was too dangerous to stay. Tourists stopped coming and their absence cut off the income that supported the work of the park. Wars have come and gone throughout history and the problems in Rwanda will eventually be settled, but the situation shows just how vulnerable a small population of animals can be when swept up in the maelstrom.

Following the war, massive refugee camps appeared in Zaire and all were starved of food and other resources. The desperate inhabitants were forced to fend for themselves. Large areas of Virunga woodlands were cut down to provide building materials and firewood. At one stage, it is thought, the deforestation took place at the rate of 325 hectares (800 acres) a day; it is doubtful that the forest will ever be allowed to regenerate. There is now a movement afoot to have the whole of the Virungas declared a World Heritage Site. This would offer no extra legal protection to the mountains, but the high-profile status should ensure that any threat or damage to the area would result in worldwide publicity and hopefully a tidal wave of political and public pressure strong enough to stop further problems.

10
THE MYTHICAL APE

APES DO NOT appear in traditional European folklore, for the simple reason that they were totally unknown in the West until comparatively recently. They obviously had a role in the mythology of their native lands, but in all cases, the legends were passed on in oral rather than written form and the majority have either been lost or not yet recorded. Some, however, have survived to reach a wider audience. In Indonesia and Malaysia many stories circulate about local women who have been raped by rampaging male orang-utans. One is very well known and tells of a sleeping woman who was woken from an afternoon nap by a 'fat monkey' or, to be more precise, an over-amorous orang-utan, who turned his attention to a human in the absence of a female of his own species. In this particular case the woman is reputed to have escaped without harm.

There are reports of orang-utans being kept as 'sex slaves' by locals with unusual tastes. Exaggeration must play a part in these anecdotes, but they may not be as outrageous as they at first appear. Young male orang-utans are excluded from mating by powerful higher-status males and this results in frustrated behaviour that may get out of hand and become redirected. With their overwhelming strength and natural propensity for using power to overcome female reluctance, lone male orang-utans may occasionally be tempted off the beaten track. The case has yet to be settled either way, but it would be unwise to reject the possibility.

Chimps and gorillas have been on the receiving end of similar slanderous stories. One report in the early nineteenth-century tells of chimps that kidnapped young ladies 'who sometimes escaped to human society after having been for years detained by their ravishers in a frightful captivity'. While there may be a possibility of a male orang-utan showing interest in human females, no one really accepts the idea that other apes are likely to behave in this way.

The popular modern view of apes has been shaped by movies, television and mass-produced books, almost all of which have chosen one of two options. Apes are either clowns for our amusement or they are rampaging monsters that create havoc before being defeated by superior humans.

It is the human appearance of apes that makes them perfect subjects for myths and legends.

One of the earliest fictional appearances of an ape came in Thomas Love Peacock's satirical novel *Melincourt* (1817). This tells the story of Sylvan Forester, a young philosopher who was convinced that apes were related to humans. He obtained an orang-utan and taught it the finer things, barring speech, that were needed to succeed in sophisticated company. Following intensive training Forester purchased a baronetcy for the ape, together with a seat in Parliament. Sir Oran Haut-Ton was readily received by the higher echelons of polite society and was in great demand for his ability as a sweet and accomplished flute player. But it was noticed that, after too much wine, Sir Oran was inclined to leap through open windows and disappear into the woods. His persistent silence in Parliament was widely respected, as this was a sure sign that he was in perpetual deep thought.

In Edgar Allan Poe's story 'The Murders in the Rue Morgue' (1841), an orang-utan was the unwitting tool that caused the death of two women, although when the film came to be made it was a gorilla that carried out the crimes, probably because the species was rather better recognized. Dr Mirakle, the obligatory mad scientist, kept the ape on public display and used it to demonstrate Darwin's theory of evolution. As a souvenir, visiting ladies were given a bracelet from which hung a small bell. At night the gorilla was turned loose to track down the unsuspecting victims by following the tinkling of their bells. They were then carried back to the doctor, who drained their bodies of the fresh blood needed for his experiments.

This mountain gorilla is sucking his thumb. Such childlike actions tempt the viewer into feeling that apes are even more like humans.

Probably the most enduring ape-related myth was dreamed up by Edgar Rice Burroughs in his Tarzan books. The plot is simple and effective. A human baby boy was abandoned in the jungle. Luckily, he was found, and was then reared by a family of apes. The boy grew into Tarzan, a man fully at home in the forest and an accepted member of his adopted species. The Tarzan story was quickly adapted by the blossoming movie industry and a seemingly endless series of low-budget films was produced showing the ape man overcoming insuperable odds to save the jungle and its residents from assorted dangers. To add an element of comedy a young chimpanzee, Cheetah, was introduced into many of the films. The animal never actually appeared in Rice Burroughs' original stories, but it was thought that a mischievous 'clown' would bring some light relief into the dramatic plot of the films. At least nine different chimps appeared in the series, each being replaced once it was too big and uncontrollable.

The early films containing Cheetah were shot entirely in and around USA studios, but as the production budgets grew, location scenes were eventually shot in Kenya. Two captive chimps were shipped over from England to play their character role in more realistic surroundings. Both animals were so frightened by the strange environment that their parts in the script had to be dramatically shortened. Faced with the real Africa, they either hid, refused to move or ran away. One was so disturbed that it could be tempted to appear in front of the camera for just long enough to hear Tarzan say, 'So long, Cheetah.'

The concept of a human baby being reared by wild animals was hardly a new basis for a story. Rudyard Kipling used the idea in his *Jungle Book* and it dates back to Roman mythology, in which Romulus and Remus were adopted by a she-wolf. Burroughs was both enterprising and lucky because he wrote the Tarzan stories at a time when Africa was unexplored, exciting and highly marketable. The unfolding melodramatic details about the exploration of the continent's interior were major news in the press and the enormously popular public lectures of the day. Close behind, exploiting the public interest, hundreds of novels were published set in steamy jungles and showing brave European explorers pitting their wits against wild animals and a savage landscape. However, it was tales of the ape boy that most caught the imagination of the public, and Tarzan has since become part of modern Western culture.

In the next century Pierre Boule introduced a new twist in his book *Planet of the Apes*, in which an earth of the future is inherited by the descendants of apes. They have mastered spoken language and finally unseated humans as the dominant species. The story goes on to tell of the struggle between the human heroes and the evil usurping apes.

The ape image that everyone immediately recalls, even if they have never seen the original film, is King Kong clinging to the top of the Empire State Building while trying to fight off puny biplanes that shoot at him from all directions. This is one of the most famous scenes in movie history, due to the fact that it employed special effects that were quite extraordinary for their time. The adventures of the 15m (50ft) gorilla were the brainchild of Mervin C. Cooper, one-time explorer and documentary film-maker. The plot told of a giant gorilla that lived on an island in the Indian Ocean, where it terrorized the locals. The animal was captured alive and shipped to New York, to be displayed as a curiosity that would make money for its captors. King Kong was scheduled as a horror/disaster film but scenes showing him tethered in a nightclub and his eventual death were handled with surprising sensitivity. Films of this genre tended to show the 'monster' as an outright baddie worthy of little sympathy, but by the end of the film, it was obvious that King Kong was to be seen as an innocent victim of human greed – a very unusual viewpoint in the popular media of the 1930s.

Non-fictional personal accounts of apes are rare before the mid-nineteenth century, but monkeys were familiar to Europeans long before that time. An entry in Samuel Pepys's diary for 24 August 1661 gives us a good idea of the general feelings about primates before the facts were revealed in more scientific terms:

By and by we are called to Sir W. Battens to see the strange creature that Captain Holmes hath brought with him from Guiny; it is a great baboone, but so much like a man in most things, that (though they say there is a Species of them) yet I cannot believe but that it is a monster got of a man and she-baboone. I do believe it already understands much english; and I am of the mind it might be taught to speak or make signs.

When apes first appeared in the West they created an uproar and became big business. At the end of the nineteenth century Wombwell's Menagerie became the proud exhibitors of the first live gorilla ever seen in Europe. It was to be a further twenty-three years before the USA saw its first living specimen. Gargantua was a huge lowland gorilla that appeared in circuses all over the USA and played to an estimated audience of 40 million people in his lifetime. He was shown by both Ringling Brothers and Barnum & Bailey from 1938 to 1949, and was the acknowledged star attraction of both. Gargantua had an appalling life in captivity and was terribly treated at times. Acid was thrown into his face at a very young age, resulting in scarring that gave him a permanently fierce grimace. Subsequent brutality at the hands of later keepers gave him a matching temperament. Copywriters capitalized on his gruesome appearance, their adverts boasting:

The World's Most Terrifying Creature, with a smirk of cruel calculation and a sadistic scowl of challenge on his bestial face. Gargantua the Great now defies civilization from behind the heavy, chilled steel bars of the strongest cage ever built.

Baby orang-utans: it is almost impossible to imagine animals that are better suited to appeal to human beings, which is why apes have such a strong pull on our emotions.

Man-like animals have always featured in fables from exotic lands and there is a school of thought that believes at least some of these must have a basis in truth. The International Society of Cryptozoology was established to examine the possible existence of species not currently recognized by science. The Secretary of the Society, J. Richard Greenwell, has listed four anthropoids of myth that he believes may actually survive in remote areas. He gives the Bigfoot or sasquatch 'a fairly high probability of existence'. There have been over 3,000 eyewitnesses who claim to have glimpsed this giant ape, around 2.2m (7ft) tall wandering around the forests of western North America. After twenty years of study, Grover Krantz, an anthropologist from Washington State University, believes the Bigfoot is a relic population of *Gigantopithecus*, Asian giant apes that are generally regarded as having been extinct for more than 300,000 years.

Similar explanations have also been attached to the Yeren of central China. Local witnesses tell of large red primates that walk about on four legs. A number of hairs have been discovered and analysed; the results showed that they belonged to no known primate species. There is speculation that the Yeren may be a form of orang-utan that still clings on in China, which was certainly part of the species' original range. In Asia there have been numerous sightings of a creature that resembles humans rather than apes. Some cryptozoologists are convinced that these are the last surviving Neanderthals, which are generally supposed to have disappeared 35,000 years ago. Recent reports tell of a group of Chinese villagers who watched a peculiar animal swim across a river; it was a creature that no one had seen before. Eyewitnesses describe the beast as

The playful nature of gorillas belies their murderous popular image.

both 'a strange black creature' that looked like a man and 'a monkey with golden hair'. The lack of agreement over the animals appearance makes identification impossible.

The most infamous man-ape of modern times is the Himalayan Yeti, but few zoologists are even willing to consider the possibility of its existence. The scanty evidence is based around strange footprints that do not fit any known animal. But as lying snow shifts and contours are altered with changes in temperature, even conventional footprints can quickly become distorted. From what we understand about ape ecology, it would be impossible for any ape-like creature to survive in the snowy wastes of the Himalayas. Winter conditions would be too prolonged and harsh, providing precious little for the animals to eat. The Yeti has to be filed away, together with mermaids and unicorns, as a romantic but purely fanciful mirage. However, it would be wrong to completely discount the possibility that unknown primates still exist. Only recently a large antelope was discovered in Vietnam. This is a well populated and thoroughly explored region, yet it was home to a donkey-sized animal that was totally unknown to science. We should not be so arrogant as to believe that all species have been found and classified.

Earlier this century several reputable books seriously discussed the existence of a curious beast known as the 'kooloo-kamba'. Paul du Chaillu, a traveller who appears briefly in the story of the discovery of gorillas, included in his book a description of a wild ape that was supposedly the offspring of the mating between a gorilla and a chimpanzee. The likelihood of this event was solemnly debated by early zoologists, but most modern researchers dismiss the theory as preposterous considering the difference in size between the two apes. No male chimpanzee could approach a female gorilla without incurring the wrath of a silverback. And, as a male gorilla is approximately four times the size of a female chimp, the anatomical difficulties and sheer terror involved would prevent successful mating. It is also worth remembering that the two are distinct species and therefore genetically incompatible for breeding purposes.

Recent images have become just as entrenched as earlier opinions. There can be few people whose impressions of chimpanzees have not been coloured by the sight of these apes, dressed in human clothes, holding saucers and drinking from cups in a parody of the English tea-party ritual. It started purely as an entertainment at London Zoo and quickly caught the imagination of the visitors who saw it. Many zoo professionals now steer away from the idea of humanizing animals in this way, preferring to show them in a more natural state. But still the tea-party idea lingers. Few visitors appreciate that only very young chimps could be used, because older ones were likely to become too excited and uncontrollable. The apes themselves probably enjoy the experience. Chimpanzees are curious creatures and the novelty of playing with clothes and crockery would have helped break the boredom of cage life. The danger lies not with the animals, but with the audience. Viewers knowing little about the true nature of chimpanzees could watch a tea-party performance and leave thinking that these cute characters were harmless and docile. Nothing could be further from the truth. Powerful images such as these must have tempted many people to buy young chimps as pets when it was still legal. Time would quickly have shown them the error of their ways.

Any well-publicized and widely accepted distortion of facts about wild animals is potentially dangerous. The film *Jaws* did untold harm as it whipped up a storm of anti-shark feeling around the world. Thousands of harmless animals were probably killed as a result. When Clint Eastwood's blockbuster *Every Which Way But Loose* was released, animal dealers and zoos everywhere were inundated by callers demanding to buy an orang-utan as a pet. The film showed a friendly orang-utan living as part of a family; it shared the house and car and even went shopping with the human hero. It is true that some orang-utans do reach this level of calmness and predictability, but finding themselves in the circumstances enjoyed by the screen ape, the majority would cause a trail of destruction that can only be imagined.

One of the most recent screen apes appears in a movie called *Congo* based on a book by Michael Crichton. It features a gorilla by the name of Amy that has mastered communication and needs to be treated as an equal. The scientific basis of the story is simply an extension of current research, although it is extremely unlikely that any ape could ever begin to approach the linguistic skills of Amy. Unfortunately, the story goes further and ends with wild gorilla-like animals killing a group of humans. And so the myth continues relentlessly on its way: apes are either harmless clowns or bloodthirsty killers.

11
INTO THE FUTURE

THE QUESTION OF the future survival of apes is clouded by one overriding problem and that is man and how he handles the environment. Humans have always been guilty of overkill when dealing with wild animals, particularly where monetary profit is involved. Education programmes are now teaching us that to maintain a long-term sustainable 'harvest', the killing must be moderated for the animals to continue providing food in the future. There are still too many examples of officially sanctioned overhunting and overlooked illegal poaching throughout the world, although the problem is being faced now and, in a few cases, controlled.

Even pollution is being tackled, albeit with mixed results. In the face of increasing public concern and subsequent political pressure, the dumping of man-made poisons is monitored and in some areas contamination has been reduced. But that is probably as a result of the danger it poses to humans as much as to wildlife. Other threats are more direct, for humans have always been keen on owning wild and potentially dangerous animals. The demand for exotic pets certainly had an impact on the wild population of those species that were regarded as collectable. International laws are being tightened to reduce the once huge trade in live specimens of endangered species and their skins. It would be naïve to believe that the business will ever be destroyed, but over the past decade it has certainly been dealt a severe blow. Popular opinion has radically changed. Once the wearing of exotic fur coats was a status symbol in Western culture, while in most places this is no longer acceptable. Today anyone wearing a genuine animal skin in public is running the risk of being socially ostracized, pelted with tomatoes or even physically attacked.

Put together, these actions are making some contribution to conservation, but from only one direction. The world's wildlife is facing another threat just as insidious and fatal as overhunting, and one that is proving far more difficult to control. Direct exploitation needs to be controlled in order for endangered species to have any chance of survival, but there seems little point in drawing up tough laws and sophisticated education projects about saving animals unless the habitat in which they live is equally well protected. The biggest single danger now facing the majority of wildlife species is destruction of habitat.

The unavoidably artificial conditions in rehabilitation camps produces unusual behaviour in orang-utans. In the wild state youngsters would never play together in large 'gangs'.

More than three-quarters of the world's animal species have evolved for life in specific environments. If that habitat is destroyed or significantly altered, these species can no longer survive. Habitats not only supply food; they also sustain critical humidities, offer suitable breeding grounds and maintain complex, deeply interdependent ecosystems where many species can live only if all the others are present. The destruction of a habitat kills its occupants as surely as a bullet in the brain. The devastation of the world's tropical rain forests has been well publicized over the past few years. However, it is of such massive and far-reaching importance that the warnings of the looming catastrophe must be repeated again and again until we all realize exactly what is happening and call a halt before we reach the point of no return. It has been estimated that the rain forest is being destroyed at the rate of 90 hectares (220 acres) each minute, and that is every single minute - day and night - of the year. Primary tropical forests covers around 11 million square km (4¼ million square miles) in total. Even this huge area cannot last long when it is being cut down at the rate of nearly 5 million hectares (12 million acres) a year. These figures are impossible to imagine, so, to make the destruction more comprehensible, we are losing rain forests at the rate of about 130,000 hectares (322,000 acres) every single day. And the resident wildlife is either wiped out or forced to find a new habitat.

From a strictly human viewpoint, apes elicit particularly strong reactions. People find them interesting and, as a result, apes have accumulated a certain amount of political and economic muscle. Universities, conservation groups and individuals all over the world are busy raising cash, producing scientific papers and creating an awareness of the great apes. And all of this is vitally important. The real key, however, is to work out exactly what can be done to guarantee a future for these species. Gorillas, orang-utans and chimpanzees have all now been officially recognized as globally endangered animals. Under the CITES (Convention on International Trade in Endangered Species of Wild Flora and Fauna) Agreement, neither live specimens nor their products (skin, bones, teeth, etc.) can be freely bought or sold.

Apes are included in Appendix 1 of the Agreement, which lists the most threatened species whose survival is to some degree affected by trade (some species are severely endangered but may not be caught up in international trade). CITES states that any commercial dealings involving these animals or their products have to be fully licensed, documented and monitored. International trade in wild animals is potentially a billion-dollar business and needs to be closely policed, as many unscrupulous dealers will ignore or bypass the law because of the huge profits at stake. CITES exemption certificates are occasionally issued for research work or captive-breeding programmes. These should be printed on special security paper, complete with embossed stamps or seals to reduce the possibility of forgery. More than 120 countries are signatories to the Agreement and

Tropical rain forest in Borneo: the destruction of this complex environment is the single biggest factor threatening the survival of many apes.

this has gone some way to reducing the number of live animals being sold as 'pets', as well as cutting down the market for trophies or souvenirs made from endangered species. The CITES Agreement is a step forward, but we have to remember that it is *just* an agreement and not an international law. Like so many good ideas worked out at a global level, CITES in practice is far less effective than it is on paper. The enforcement of the legal aspects such as licensing and the control of smuggling is left to individual country signatories, and some are more conscientious than others.

Apes have long been popular as pets, but only while they are still young and cuddly. They soon outgrow their cute appeal and then are no longer welcome houseguests. Before the CITES agreement, captive apes could just be sold to zoos or circuses, or even killed for the value of their skins and other trophies. Obviously this still goes on, but today it is far more difficult for an animal dealer to make money out of an illegal ape than it once was. Through a tightening up of rules and a greater public awareness, many pet apes are being confiscated by the relevant authorities, but they are proving to be a major problem.

Few confiscated animals have been captive-bred and the vast majority would have been captured at a very early age. After just two or three years in captivity, they are no longer wild animals, but have become something completely different. Instead of being experienced juveniles with a deeply ingrained knowledge of their environment, along with its food supplies and dangers, humanized apes become dependent, semi-domestic creatures with little fear of people and no ability to fend for themselves in their natural environment.

Orang-utans have long been popular pets in Asia. Although it is illegal for a private citizen to own one, a few slip through the net and every year a number are confiscated and taken into care. Both Indonesia and Malaysia have efficient, well-run rehabilitation centres in protected forests and national parks where permanent staff are on hand to encourage the young orang-utans to slide gradually back into forest life. However, things never work out quite like that. Wild orang-utans feed on more than 400 known plants. They might take only the fruit or roots of some species and these may be edible only at certain times of the year. Young orang-utans learn where and how to find food by carefully watching their mother feed. The long dependency period ensures that juveniles have a thorough grounding in the skills they need to survive.

If a baby orang-utan is forcibly taken from the forest at six months old and does not return for the next four or five years, a critical stage of its education has been lost, a stage that can never be replaced. The staff at rehabilitation centres may be dedicated, but it would be completely impossible for them to take each youngster out into the forest and demonstrate what food to eat and exactly where to find it. They would not have the time or the knowledge to carry out this massive task, and a large number of newly confiscated orang-utans are too old to learn everything.

The question of rehabilitation and reintroduction is a thorny one. Not all zoologists agree about the worth or long-term viability of such programmes, for the problems are numerous and complicated. First, orang-utans are, physiologically, so similar to humans that they are vulnerable to many of the diseases that affect us. In the wild there would be little opportunity for infections to travel freely between humans and orang-utans, but in captivity the animal may be passed from hand to hand, before being confiscated, moved, examined and finally transported to a release site. The unfortunate animal would have had any number of chances to pick up a cocktail of potentially lethal bugs which might then be transferred to the local wild population. Animals are always checked by vets before release, but some infections can lie dormant for years before symptoms become visible. Released captive animals usually retain a tolerance and even a curiosity about humans, climbing up legs and dropping on to shoulders from overhanging branches. Rehabilitation centres attract tourists from all over the world, any of whom may introduce infections which eventually could be passed to wild orang-utans that would never normally come anywhere near a human.

On my first introduction to rehabilitated orang-utans at Camp Leakey in Tanjung Puting National Park, in Indonesian Borneo, I was so keen to photograph a tiny three-year-old female that I completely failed to notice an older male sneaking up behind. His long arm snaked into my open rucksack and grabbed a Pentax camera. Orang-utans have a unique ability of looking slow and clumsy while moving incredibly fast. The male was up a tree and out of reach, clutching his booty, within seconds. A few cursory sniffs and shakes convinced him that the small black object was not particularly interesting, so he began to beat it half-heartedly against a thick branch. Orang-utans are powerful creatures even at six years old. The sickening crunches indicated that the camera could not take this treatment for long. I finally got it back by tempting the animal down with a banana, a favourite food of orang-utans. Surprisingly, the camera was still in working order, with no serious damage, but the episode taught me that rehabilitated orang-utans are very different from their wild cousins.

It is impossible to take a newly confiscated orang-utan and dump it in a rain forest – a house-trained poodle would have about as much chance of survival. These animals have been dependent on humans for too long and they must be very slowly introduced back to their native environment. For the first few months the orang-utans stay at base camp, where rangers feed them on milk and bananas. As they grow in confidence, food is left deeper in the forest, encouraging them to explore and venture further afield. Obviously they experiment and learn to eat local food, particularly fruit and leaves, but their feeding knowledge is superficial compared to their wild counterparts. So their diet must be supplemented by food from the rangers. Within weeks, the highly intelligent orang-utans have worked out that here is the best of all worlds. They have the freedom of the forest and can come and go as they wish, while still being fed and entertained by humans.

The lure of food, particularly bananas, keeps them close to base camp. They may disappear into the forest for days at a time but always return. Some individuals have been centred around a rehabilitation camp for decades. On reaching sexual maturity, they mate with other camp animals or even wild orang-utans. Female rehabilitants are certain to introduce their offspring into this artificially dependent world. The youngsters too become acclimatized to the presence of humans and the non-stop supply of food, and thus institutionalization is passed on to the next generation. Rehabilitation camps have been described as being more like semi-wild breeding programmes rather than anything else.

These strange conditions result in major behavioural changes. A wild young orang-utan would be brought up by its mother and, as a pair, the two would rarely come into close contact with other individuals. This fits in with the solitary lifestyle that the animal will later experience as an adult. In and around rehabilitation camps there may be thirty or forty juveniles of all ages. They feed, sleep and play together – something that would never happen under normal conditions. After a while, they form small, loosely structured 'gangs' that resemble chimpanzees' social groups rather than orang-utans'. The feeding patterns and the very presence of ten apes have a markedly different effect than just one. Together they completely strip trees and drive off other wildlife, particularly the wary gibbons and proboscis monkeys. Camp orang-utans become far more sociable than they ever would have done in the forest. No one truly understands how this early learning governs their subsequent behaviour, but it is certainly a poor apprenticeship for an animal that is genetically programmed for a solitary life.

There is now an alternative approach to rehabilitation, in which contact with humans is kept to a bare minimum. Before anything else, newly arrived orang-utans are kept in quarantine for six months. This buffer period helps park staff to spot and treat human diseases that may be taken into the wild and to sort out problems such as vitamin and mineral deficiencies the apes may have suffered in captivity. This time helps the orang-utans to acclimatize. At first they are more human than ape in behaviour; it takes some time for them to become accustomed to their own species. Each group intentionally contains animals in sex ratios that are found in the wild. After the quarantine period, the rehabilitants are taken out deep into a forest where there are no resident orang-utans to compete with. They are released far from tourist trails, keeping contact with humans to a minimum. They are fed for three months by rangers while they learn to find and identify their own food plants. Early results indicate that this approach is more successful than other programmes, producing a remarkable survival rate of 95 per cent in the first two years.

But more work needs to be carried out to ensure that the orang-utans actually survive for a reasonable time in the wild and go on to breed successfully. It is this last point that will prove the project's true viability. While the rehabilitated animals can never be regarded as truly wild, with luck their young could be.

Ecologically, reintroduction is useful only when the existing wild population is very low. Released animals can then be used to top up their numbers and introduce new genes. Where orang-utans occur in healthy ecosystems they are usually at saturation level and the habitat is not capable of supporting more. If there are no orang-utans in an apparently suitable forest, there is probably a good reason for their absence, even though we can't see why.

When a young orang-utan arrives at a centre, often little is known about its origins. Young Bornean and Sumatran orang-utans look identical and in the illegal animal trade apes are supplied from both islands. Although the two subspecies can interbreed, conservationists feel that this should be avoided wherever possible. If a large number of the 'wrong' orang-utans were released into either island, they could affect the genetic make-up of the resident population. This must have already happened, and some biologists regard the mixing of strains as a disaster, because effectively the two subspecies will eventually disappear through interbreeding. Some primatologists go even further and maintain that the risk of merging different genes of orang-utans from eastern and western Borneo is enough to question the continued existence of rehabilitation centres.

So, if rehabilitation is not the answer, what can be done with the illegal orang-utans? Few zoos want them, as they are expensive to keep, and they live an awfully long time in captivity, so the market demand is low. They could, of course, be humanely destroyed, but this would be unacceptable to everyone. It has been suggested that camp orang-utans should be fitted with long-term contraceptive implants. This would prevent the animals breeding and passing on their atypical behaviour to their offspring. Such a system would also allow reintroduced orang-utans to live out the remainder of their lives under pleasant conditions. Rehabilitation centres may not be the perfect answer to humanized orang-utans, but they may be the only practical solution to a highly complicated problem.

The paradoxes posed by rehabilitation also apply to the idea of captive-breeding-and-release programmes. The technique has been used for centuries by gamekeepers, whose task it is to provide the maximum bag for hunters paying to shoot pheasants and partridges on organized estates. Zoo-bred animals like the ne-ne, or Hawaiian goose, have been reintroduced to their native habitat with great success. Such projects work because each of these animals is a non-specialized feeder that does not need a deep knowledge of its habitat to survive. Theoretically, it would be possible to release any species that eats readily available food and does not need to fit into complex social groups. Apes are the exact antithesis of this, though.

When they arrive at rehabilitation camps, newly confiscated orang-utans are often withdrawn and badly frightened.

move on. The remaining land is abandoned or turned over to agriculture, where it will provide a meagre living at best. The end result is cash today and absolutely nothing tomorrow.

Such operations run on a fast-turnaround system. Big trees, the source of high-profit timber, are taken first. Smaller trees, which are not worth processing, are burned as fuel or pulped for the production of reconstituted wood products such as chipboard. Tropical rain forests are unique in so many ways. They must not be treated like arable land, which can be planted and harvested year after year. In most habitats, the earth itself is the storehouse of nutrition, minerals and the other elements that make life possible. Savanna grass takes what it needs from the earth through the root system. The plant grows quickly in spring and dies down in winter. The elements return to the soil ready to feed new plants next year. It is a very rapid cycle.

In mature rain forests the colossal trees take up so much from the soil that they become the storehouse of nutrition and minerals. A small percentage is annually given back to the earth in the form of dropped fruit and leaves, but the bulk is incorporated into the trees' living tissue and there it stays until one of the giants dies. Decomposition then does its job and the elements slowly return to the soil. With a sudden opening in the high canopy, light floods down to the forest floor and stimulates the growth of the tiny saplings that would normally die in the gloom or be stunted through competition. A race begins to reach the life-giving light and soon one young tree dominates and grows to fill the gap. Its expanding root system gets valuable nourishment that is slowly released from the corpse of its deceased predecessor. The canopy is closed surprisingly quickly and the forest floor soon returns to its permanent twilight state. The complete cycle is agonizingly slow, as these trees can live for centuries and keep nutrients locked away from other life forms. However, it works well, for the deaths of the great giants are not synchronized but are occurring all the time around a forest. The cycle may be unhurried but inevitably the vital elements revert to the soil so that life begins again. Catastrophe occurs when industrial man arrives to take away the mature trees, because the bulldozers and lorries do not just remove the wood, they steal the very lifeblood of the forest. Through practice, the industry is swift and efficient, the camp never stays in one place long before moving on to the next stand of trees.

Once the loggers have finished their task and moved on to the next part of the forest, foraging domestic animals such as goats or cattle often take their place. Quickly eating the remaining edible plants, they act as efficient land scourers. The sparse vegetation supports them for only a short time and once there is nothing left for them to eat, they are shifted to clear the next parcel of demolished woodland. Agricultural cultivation is the next step, but there is so little nutrition in the soil that crops can be grown for only two or three years before the site is exhausted. Its inherited richness disappeared with the trees and the ground is now so impoverished that even subsistence farming is impossible. In five years a rain forest, probably the richest environment on earth, can be transformed into a near-sterile scrubland.

It was a sad day for the local forests and wildlife when gold and diamonds were discovered in Borneo. The immediate economic advantages are obvious, but the long-term drawbacks will be felt for years to come. Many of the new mines are open-cast for simplicity. First, the trees are felled and the soil is scraped away. Instead of using expensive digging machines to reach their goal, where possible operators are shifting soil with water pumped from nearby rivers. It is blasted through high-pressure hoses, producing a relentless barrage of water that turns the soil into mud and washes it away in countless new streams, leaving the minerals in place but exposed. The technique is quick and cheap, but it results in terrible wounds to the forests and clogs up the waterways with thick mud that kills the fish and later settles as silt that blocks the rivers.

We simply cannot view tropical forests in the same light as other natural resources. Coal, oil, diamonds and the rest of our planet's commodities are the remnants of long-dead ecosystems; once deposits are spent we might be deeply inconvenienced and extremely depressed, but life will still go on. Rain forests are not fossilized remains; they are the irreplaceable backbone of a vibrant habitat that supports possibly half of all the living organisms on the planet. If the loss of all this doesn't strike horror into the hearts of the human race, then perhaps the potential risk to our own safety might.

Much has been written about the greenhouse effect and subsequent global warming, which may be just one result of our mistreatment of the environment. Without dwelling on the biochemical details, part of the problem is the large-scale destruction of tropical forests. For these have been called the 'lungs of the Earth' because they manufacture so much of our oxygen. Despite painstaking research throughout the world, we still do not really understand the full implications of the greenhouse effect. Scientists can make educated guesses about the nature of the repercussions, but that is all they do until something actually happens, and by then it will be too late. We could all be wiped out by a massive rise in the incidence of skin cancer, or drowned because the polar ice-caps have melted, causing the sea level to reach unknown heights. None of the suggested scenarios is very attractive.

It may be possible to artificially create and maintain rain forests for the benefit of both big business and wildlife, but at the moment we just don't know how to do it. And, to be blunt, there is no real commercial incentive to sink huge amounts of money into such research projects. With typical human short-sightedness we stumble on, crossing bridges only as we reach them. We treat rain forests just as we treat oil – easily available *now*, and tomorrow is another day.

Unfortunately, many results of rain-forest destruction are either delayed or felt elsewhere, making them easier to ignore. 'Downstream effects', as they are called, de-personalize the consequences, making them remote and less threatening to those responsible. We can all get worked up over hazards on our own doorstep when family and friends are at risk, but we can overlook dangers that may be felt in ten years' time in another part of the world – particularly when the causes are those actions that put food on the table now and keep a roof over the family's head. The developed world is in a safe position to demand a cessation in large-scale logging because our economic success does not depend upon it.

Priorities alter according to personal perspective. From the security of a city that relies on, say, banking or car production for income, we can articulately and forcefully argue the case for protected forests. Facing bankruptcy, an underfinanced government with a human population demanding its share of the good life has a different viewpoint about the value of tropical woodlands. We should all remember that it is the West that offers the biggest market for products made of tropical hardwood, so our moral righteousness is built on shaky foundations – particularly when we point an accusing finger at countries where forests are one of the few economically valuable assets. Disregarding the effects of commercial logging, tropical forests are being eroded as a result of the sheer numbers of people who now inhabit our planet. Each one requires food, shelter and fuel, and in almost every culture people have increasing expectations of life. Once we needed just a simple hut in which to shelter from the weather, together with enough food to fill our stomachs for the day. Now we want electricity, televisions, refrigerators – the list is endless. The production of such things has little direct effect on the forests, but indirectly consumerism may be the catalyst for their destruction. With higher aspirations comes a need for more money to pay for the new lifestyle. Instead of wanting just the essentials for day-to-day living, we now must earn enough to buy the luxuries. Today it might take the proceeds from 30 hectares (74 acres) of forest timber to keep a prosperous Indonesian family for twelve months, while 50 years ago it would have taken less than 1 hectare (2½ acres) to supply all their needs. Because the developed world is the driving force behind the large-scale production of consumer goods, we have to find someone to buy them. Aggressive advertising is creating a demand for commodities that are partially being bought with profits from the sale of forest products. The West may not be wielding the axe, but it is certainly paying the axeman's wages.

Developing countries may have the will and desire to conserve their forests but they must be practical about the problems they face. Politics has been described as the 'art of the possible' and this can clearly be seen in Indonesia. The country is made up of countless islands ranging from Java, a huge land-mass, down to tiny uninhabited rocks. The human population is nearly 200 million and growing rapidly. Well over half of all Indonesians live on Java, an island with one of the highest population densities in the world. Every possible corner is cultivated to provide food for people and their livestock. To the west and north lie Sumatra and Kalimantan respectively, Indonesian areas with tiny human populations compared to Java. It seemed obvious to move people from overcrowded Java to the near-empty lands lying just across the water. So the government introduced an aided transmigration programme that was designed to spread the population out a little more evenly. It hasn't worked quite as planned.

The transmigration project relies on volunteers pulling up roots and starting afresh in a new area. Unfortunately, good farming land was already taken, even on the thinly populated islands of Borneo and Sumatra. Newly arrived immigrants were allocated plots that were less than ideal, and that included rain forest. The majority of those who apply to the programme are either urban dwellers tired of poverty and life in shanty towns, or homeless. Through no fault of their own, neither group has the experience to take a piece of virgin land and cultivate it efficiently. They have an urgent need to stake a claim in their new land and the quickest way to earn money is to sell the standing timber, and so the forest soon disappears. Those who have a real chance of working the land properly, the farmers and foresters, are already employed and have no need to move, so they remain in their homes on Java.

Most of the 'new farmers' exist at subsistence level. They find the soil poor, are too far away from major markets and lack the skills to grow crops or tend animals. According to the Indonesian government's Second Five Year Plan, the target was to move a little over 5 million people from Java on to other islands. Just a tiny fraction of this figure would be enough to disrupt the ecosystem of a forest and its wildlife, including the orang-utans.

I am trying hard not to be pessimistic about the long-term fate of the great apes, but a gloomy realism stops any feeling of optimism. We have to acknowledge that these animals – and most other tropical-forest inhabitants – face a less than rosy future. Time and the majority of humans are not on their side. The apes can be saved, but only if we do something now. Unless we conserve the forests instead of destroying them, our grandchildren may never have the chance of seeing a live gorilla or orang-utan outside a zoo.

Extinction is a word that is tossed around every day by the media, and that's good, for we all need to be reminded about the potential consequences of our actions, but I feel the word is slightly misleading. Animal species died out long before humans came along and interfered; extinction is part of the natural cycle of evolution. In the modern world, most cases of extinction are not natural; they are man-made. If the apes should die out, it will be because we have killed them. And the correct word to use then would be genocide.

BIBLIOGRAPHY

Diamond, J., *The Rise and Fall of the Third Chimpanzee*, Vintage, 1992

Dixson, A. F., *Natural History of the Gorilla*, Columbia University Press, 1981

Fossey, Dian, *Gorillas in the Mist*, Houghton Mifflin, 1993

Galdikas, Birute, *Reflections of Eden*, Victor Gollancz, 1995

Goodall, Jane, *In the Shadow of Man*, Collins, 1971

– *The Chimpanzees of Gombe: Patterns of Behaviour*, Harvard University Press, 1986

– *Through a Window*, Weidenfeld & Nicolson, 1990

Heltne, P. G. and L. A. Marquardt (ed.) *Understanding Chimpanzees*, Harvard University Press, 1989

Kano, T., *The Last Ape*, Stanford University Press, 1992

Lemmon, Tess, *Chimpanzees*, Whittet Books, 1994

Montgomery, S., *Walking with the Great Apes: Jane Goodall, Dian Fossey, Birute Galdikas*, Houghton Mifflin, 1991

Schaller, G., *The Mountain Gorilla*, Chicago University Press, 1976

Schwartz, J., *Orang-utan Biology*, Oxford University Press, 1988

Tuttle, R., *Apes of the World*, Noyes Publications, 1986

Wallman, J., *Aping Language*, Cambridge University Press, 1992

USEFUL ORGANIZATIONS

Bonobo Protection and Conservation Fund
c/o Dr Rose Sevcik
Georgia State University Language Research Center
3401 Panthersville Road,
Decatur, GA 30034
USA

Committee for the Conservation and Care of Chimpanzees
3819 48th Street NW
Washington, DC 20016
USA

Fauna and Flora Preservation Society
1 Kensington Gore
London SW7 2AR
(Co-ordinates and runs apes-related conservation projects)

Dian Fossey Gorilla Fund
110 Primrose Hill
London NW1 8JA
(Set up specifically to aid mountain gorillas)

Jane Goodall Institute
15 Clarendon Park
Lymington
Hants SO41 8AX
(Set up to aid common chimpanzees)

International Primate Protection League
116 Judd Street
London WC1H 9NS

Orangutan Foundation
7 Kent Terrace
London NW1 4RP

Worldwide Fund for Nature
Panda House
11–13 Ockford Road
Godalming
Surrey GU7 1QU
(Co-ordinates and runs apes-related conservation projects)

INDEX

Page numbers in *italics* refer to illustrations.

They must have an intimate knowledge of both the habitat and its food supplies, as well as fitting in amicably with the other individuals that make up the family groups. There are rehabilitation experiments for chimpanzees and gorillas but both are facing difficulties similar to those encountered with the orang-utan projects.

From simply being places of mass entertainment in the nineteenth century, modern zoos have become centres for conservation and education — or at least the more enlightened ones have. It was once thought that by breeding endangered animals in captivity we were providing a protected gene pool, a safety net for the future should the animals become extinct. There was always the thought that they might be released into the wild. Now, however, we realize we are a long way from understanding how to achieve this. Staggering advances have been made in the techniques of breeding rare animals in zoos, but releasing many of them is still proving to be a problem. Of course, this doesn't necessarily mean that it can never be done, only that we shouldn't fall into the trap of seeing captive animals as a long-term solution to depopulation in the wild. One day it may well be possible to successfully release zoo-bred apes into a forest, but not today. It is safer to utilize zoos as a way of studying animals at close quarters, enabling us to understand more about their biology and behaviour in the hope that we can help those left in the wild.

The dollar-earning potential of apes as a tourist resource is becoming increasingly important to many of the countries in which they live, and these tend to have a relatively low average income. As a result, some governments now realize that anything bringing in foreign currency is an asset to be nurtured and promoted. Unfortunately, cash can be earned in many ways and, due to the nature of the environment, ape tourism often coincides with a healthy international market for hardwoods. Orang-utans are particularly susceptible to this danger, because they inhabit forests made up almost exclusively of the ancient hardwood trees that are so much in demand in the production of high-quality furniture.

When balancing books, many countries, even in the so-called developed world, feel they cannot afford to take the long-term view. They have to look at the economics of the present, juggling the figures to pay off loans while the country just ticks over. Keeping their heads above water today is a task that borders on the impossible, so tomorrow is but a remote problem. Tourism is increasingly important, but it earns money slowly and steadily, while profits from timber sales are high and immediate. Forestry involves manpower, which is cheap and readily available, but little capital investment, and, even better, it brings instant cash. While it is possible to run commercial logging programmes in hardwood forests, long-term sustainability requires efficient management and planning to ensure a new crop of trees every year. Examples of this happening in practice are extremely rare and are known only in temperate zones. The reality in tropical forests is that loggers move in, take what they can and

Zoo apes can still sometimes be found in small concrete prisons that result in both physical and mental deterioration.